NATURAL DISASTER RESEARCH, PREDICTION AND MITIGATION

SANDY RECOVERY IMPROVEMENT ACT AND SUPPLEMENTARY FUNDING FOR DISASTER RELIEF

NATURAL DISASTER RESEARCH, PREDICTION AND MITIGATION

Additional books in this series can be found on Nova's website under the Series tab.

Additional E-books in this series can be found on Nova's website under the E-book tab.

LAWS AND LEGISLATION

Additional books in this series can be found on Nova's website under the Series tab.

Additional E-books in this series can be found on Nova's website under the E-book tab.

NATURAL DISASTER RESEARCH, PREDICTION AND MITIGATION

SANDY RECOVERY IMPROVEMENT ACT AND SUPPLEMENTARY FUNDING FOR DISASTER RELIEF

MAGNUS HASSAN
EDITOR

New York

Library of Congress Cataloging-in-Publication Data

ISBN: 978-1-62618-770-2

Published by Nova Science Publishers, Inc. † New York

CONTENTS

Preface		**vii**
Chapter 1	Analysis of the Sandy Recovery Improvement Act of 2013 *Jared T. Brown, Francis X. McCarthy and Edward C. Liu*	**1**
Chapter 2	FY2013 Supplemental Funding for Disaster Relief *William L. Painter and Jared T. Brown*	**39**
Index		**95**

PREFACE

Given its size, strength, location and the enormity of its impact, Hurricane Sandy was a storm of historic proportions and importance. The damage inflicted by this huge storm rivaled the damages of the Gulf Coast storms during the summer of 2005, particularly due to the population density and overall development of the areas that received the brunt of the storm's impact. As with other major natural disaster events, members of Congress responded to Hurricane Sandy by holding a series of hearings, visiting the affected region, and introducing and acting on legislation. This book examines the potential effects of the Sandy Recovery Improvement Act of 2013 on disaster assistance procedures and programs. Part of the legislative intent of the Sandy Recovery Improvement Act is to streamline administrative procedures and improve the effectiveness of several disaster assistance programs authorized by the Stafford Act.

Chapter 1 - Hurricane Sandy caused extensive human suffering and damage to public and private property. In response to this catastrophic event, Congress considered legislation to provide supplemental appropriations to federal disaster assistance programs. In addition, Congress considered revisions to the Robert T. Stafford Disaster Relief and Emergency Assistance Act (the Stafford Act, P.L. 93-288 as amended), which is the primary source of authorities for disaster assistance programs for the Federal Emergency Management Agency (FEMA). As a result, Congress passed the Sandy Recovery Improvement Act of 2013, which was included as Division B of the Disaster Relief Appropriations Act, 2013 (P.L. 113-2). Division A of P.L. 113-2 provided a $50.7 billion package of disaster assistance largely focused on responding to Hurricane Sandy. Additionally, Congress increased the National

Flood Insurance Program's borrowing authority by $9.7 billion (from $20.725 billion to $30.425 billion) (P.L. 113-1).

This report analyzes the provisions of the Sandy Recovery Improvement Act of 2013. In general, these provisions amend the Stafford Act with a stated goal of improving the efficiency and quality of disaster assistance provided by FEMA. Briefly, the amendments to the Stafford Act include:

- Establishing a new set of alternative procedures for administering the Public Assistance Program, which provides assistance for debris removal and the repair and restoration of eligible facilities (Section 1102 of the Sandy Recovery Improvement Act of 2013);
- Authorizing FEMA to enter into agreements with private owners of multi-family rental properties to expand post-disaster housing resources (Section 1103);
- Revising the administration of the Hazard Mitigation Grant Program, to include a possible advancement of 25% of grant funds (Section 1104);
- Directing the establishment of alternative dispute resolution procedures (including binding arbitration), building on FEMA's current appeals process, to resolve federal and state disagreements on costs and eligibility questions (Section 1105);
- Directing the creation of a joint process for environmental and historical review for disaster recovery projects with the goal of increasing the speed of the process (Section 1106);
- Directing FEMA to study, and report to Congress, whether it is appropriate to increase the dollar size of "small projects" eligible for simplified procedures (Section 1107);
- Including child care as an eligible expense under the "other needs assistance" provided in certain disasters (Section 1108(a));
- Specifically authorizing the reimbursement of the base wages of government employees providing emergency work under certain circumstances (Section 1108(b));
- Directing FEMA to update the factors considered when assessing the need for Individual Assistance in the declaration process (Section 1109);
- Authorizing the chief executive of a tribal government to directly request disaster or emergency declarations from the President, much as a governor can for a state (Section 1110); and

- Directing FEMA to create a comprehensive national strategy for reducing the cost of future disasters (Section 1111).

Prospectively, the changes in law apply to disasters declared on or after the date of enactment, January 29, 2013. Further, support can be found in the text and legislative history of the bill for applying at least some of these amendments retrospectively to Hurricane Sandy-related disaster declarations. However, it is less clear whether, and to what extent, some of these revisions will apply to disasters declared before Hurricane Sandy.

Chapter 2 - On January 29, 2013, the Disaster Relief Appropriations Act, 2013, a $50.7 billion package of disaster assistance largely focused on responding to Hurricane Sandy, was enacted as P.L. 113-2.

In late October 2012, Hurricane Sandy impacted a wide swath of the East Coast of the United States, resulting in more than 120 deaths and the major disaster declarations for 12 states plus the District of Columbia. The Administration submitted a request to Congress on December 7, 2012, for $60.4 billion in supplemental funding and legislative provisions to address both the immediate losses and damages from Hurricane Sandy, as well as to mitigate the damage from future disasters in the impacted region.

On January 15, 2013, the House of Representatives passed H.R. 152, the Disaster Relief Appropriations Act, 2013. This bill included $50.7 billion in disaster assistance. This was the third piece of disaster legislation considered by the House in the 113th Congress. H.R. 41, which passed the House and Senate on January 4, 2013 and was signed into law two days later as P.L. 113-1, provided $9.7 billion in additional borrowing authority for the National Flood Insurance Program. On January 14, the House passed H.R. 219, legislation making changes to disaster assistance programs. The rule for consideration of H.R. 152 combined the text of H.R. 219 with H.R. 152 upon its engrossment, to send them to the Senate as a single package.

The Senate passed H.R. 152 unchanged on January 28, 2013 by a vote of 62-36, and it was signed into law as P.L. 113-2 the next day.

H.R. 152 was not the initial legislative response to the storm. In the 112th Congress, the Senate passed a separate package of disaster assistance totaling $60.4 billion, as well as several legislative provisions reforming federal disaster programs. While appropriations legislation generally originates in the House of Representatives, the Senate chose to act on the Administration's request first by amending an existing piece of House-passed appropriations legislation—H.R. 1. This passed the Senate December 28, 2012, by a vote of

62-32. The House did not act on the legislation before the end of the 112th Congress.

This report analyzes the Administration's request, the initial Senate position from the 112th Congress, and H.R. 152, the legislative package developed in the House that was ultimately enacted as P.L. 113-2. It includes information on legislative provisions as well as funding levels.

In: Sandy Recovery Improvement Act …
Editor: Magnus Hassan
ISBN: 978-1-62618-770-2

Chapter 1

ANALYSIS OF THE SANDY RECOVERY IMPROVEMENT ACT OF 2013*

Jared T. Brown, Francis X. McCarthy and Edward C. Liu

SUMMARY

Hurricane Sandy caused extensive human suffering and damage to public and private property. In response to this catastrophic event, Congress considered legislation to provide supplemental appropriations to federal disaster assistance programs. In addition, Congress considered revisions to the Robert T. Stafford Disaster Relief and Emergency Assistance Act (the Stafford Act, P.L. 93-288 as amended), which is the primary source of authorities for disaster assistance programs for the Federal Emergency Management Agency (FEMA). As a result, Congress passed the Sandy Recovery Improvement Act of 2013, which was included as Division B of the Disaster Relief Appropriations Act, 2013 (P.L. 113-2). Division A of P.L. 113-2 provided a $50.7 billion package of disaster assistance largely focused on responding to Hurricane Sandy. Additionally, Congress increased the National Flood Insurance Program's borrowing authority by $9.7 billion (from $20.725 billion to $30.425 billion) (P.L. 113-1).

* This is an edited, reformatted and augmented version of the Congressional Research Service Publication, CRS Report for Congress R42991, dated March 11, 2013.

This report analyzes the provisions of the Sandy Recovery Improvement Act of 2013. In general, these provisions amend the Stafford Act with a stated goal of improving the efficiency and quality of disaster assistance provided by FEMA. Briefly, the amendments to the Stafford Act include:

- Establishing a new set of alternative procedures for administering the Public Assistance Program, which provides assistance for debris removal and the repair and restoration of eligible facilities (Section 1102 of the Sandy Recovery Improvement Act of 2013);
- Authorizing FEMA to enter into agreements with private owners of multi-family rental properties to expand post-disaster housing resources (Section 1103);
- Revising the administration of the Hazard Mitigation Grant Program, to include a possible advancement of 25% of grant funds (Section 1104);
- Directing the establishment of alternative dispute resolution procedures (including binding arbitration), building on FEMA's current appeals process, to resolve federal and state disagreements on costs and eligibility questions (Section 1105);
- Directing the creation of a joint process for environmental and historical review for disaster recovery projects with the goal of increasing the speed of the process (Section 1106);
- Directing FEMA to study, and report to Congress, whether it is appropriate to increase the dollar size of "small projects" eligible for simplified procedures (Section 1107);
- Including child care as an eligible expense under the "other needs assistance" provided in certain disasters (Section 1108(a));
- Specifically authorizing the reimbursement of the base wages of government employees providing emergency work under certain circumstances (Section 1108(b));
- Directing FEMA to update the factors considered when assessing the need for Individual Assistance in the declaration process (Section 1109);
- Authorizing the chief executive of a tribal government to directly request disaster or emergency declarations from the President, much as a governor can for a state (Section 1110); and
- Directing FEMA to create a comprehensive national strategy for reducing the cost of future disasters (Section 1111).

Prospectively, the changes in law apply to disasters declared on or after the date of enactment, January 29, 2013. Further, support can be found in the text and legislative history of the bill for applying at least some of these amendments retrospectively to Hurricane Sandy-related disaster declarations. However, it is less clear whether, and to what extent, some of these revisions will apply to disasters declared before Hurricane Sandy.

INTRODUCTION

Given its size, strength, location and the enormity of its impact, Hurricane Sandy was a storm of historic proportions and importance. The damage inflicted by this huge storm rivaled the damages of the Gulf Coast storms during the summer of 2005, particularly due to the population density and overall development of the areas that received the brunt of the storm's impact. As with other major natural disaster events, Members of Congress responded to Hurricane Sandy by holding a series of hearings, visiting the affected region, and introducing and acting on legislation.

In addition to evaluating the need for supplemental appropriations in response to this catastrophic disaster, the 112th and 113th Congresses considered reforming provisions of the Robert T. Stafford Disaster Relief and Emergency Assistance Act (the Stafford Act, P.L. 93-288 as amended).[1] Generally, concerns were raised that the recovery from Hurricane Sandy would be plagued by perceived delays and bureaucratic burdens that inhibited the recovery following Hurricane Katrina. The Sandy Recovery Improvement Act of 2013, as passed as Division B of P.L. 113-2, the Disaster Relief Appropriations Act, 2013, revises the Stafford Act as a result of these concerns as well as other considerations. Division A of P.L. 113-2 provided a $50.7 billion package of disaster assistance largely focused on responding to Hurricane Sandy. Additionally, Congress increased the National Flood Insurance Program's borrowing authority by $9.7 billion, from $20.725 billion to $30.425 billion.

This report examines the potential effect of the Sandy Recovery Improvement Act of 2013 on disaster assistance procedures and programs. Part of the legislative intent of the Sandy Recovery Improvement Act of 2013 is to streamline administrative procedures and improve the effectiveness of several disaster assistance programs authorized by the Stafford Act, namely the Public Assistance Program, the Individual Assistance Program, and the Hazard Mitigation Grant Program.[2] The law reforms the Stafford Act in part by reauthorizing several "pilot" programs established previously in Title VI of P.L. 109-295, the Post-Katrina Emergency Management Reform Act of 2006 (PKEMRA).[3]

P.L. 113-2 also authorizes tribes to request a declaration of an emergency or major disaster, as was previously only allowed for states. It also directs FEMA to establish procedures for arbitrating disputes relating to disaster assistance, and directs FEMA to create a national strategy for mitigating the cost of future disasters.

In the context of overall supplemental funding for disaster relief being debated by the Congress, the disaster assistance programs modified by the law are funded by the Disaster Relief Fund (DRF).[4] More detailed explanations of the modifications enacted by the law are provided below.

LEGISLATIVE HISTORY

On December 28, 2012, near the end of the 112th Congress, the Senate approved H.R. 1, the Disaster Relief Appropriations Act, 2013. The bill would have provided $60.41 billion in supplemental appropriations for disaster assistance and mitigation, and included a suite of legislative provisions that included reforms to several disaster assistance authorities. Included within the bill was the Disaster Recovery Act of 2012 (provided as a general provision, Section 609, to the Homeland Security title). No further action was taken on this bill during the 112th Congress.

On January 14, 2013, the House passed H.R. 219, the Sandy Recovery Improvement Act of 2013.

This bill had many similarities to the Disaster Recovery Act of 2012, as passed by the Senate in the 112th Congress, including many provisions that were identical or very similar to each other. For more on these similarities, see the **Appendix**.

Pursuant to the provisions of H.Res. 23, in the engrossment of H.R. 152, the Disaster Relief Appropriations Act, 2013, the text of H.R. 219 as passed by the House was added as new matter at the end of H.R. 152 (Division B—Sandy Recovery Improvement Act of 2013). H.R. 152 was passed by the House on January 15, 2013, and received by the Senate on January 22, 2013. The Senate passed H.R. 152 unchanged on January 28, 2013, and it was signed into law as P.L. 113-2 the next day.

EXPLANATION OF PROVISIONS

The following sections of the report examine the effect of the Sandy Recovery Improvement Act of 2013 on disaster assistance procedures and programs. The provisions are not analyzed in numerical order, but rather consolidated by the program or policy subject affected by the provisions.

Tribal Requests for a Major Disaster or Emergency Declaration under the Stafford Act

Section 1110 of the Sandy Recovery Improvement Act of 2013 amends Sections 401 and 501 of the Stafford Act which contain the procedures for requesting types of disaster declarations.[5] Previously, tribal groups were treated as local governments and thus not permitted to directly request disaster declarations from the federal government. As with local governments, the tribes were dependent on a request being made by the governor of the state where their territory is located.[6]

Tribal governments argued that the previous provision in the Stafford Act undermined their independence and sovereignty. This change in declaration policy for tribal groups had been sought by FEMA to strengthen its government-to-government relationships with tribal groups and improve emergency management in those areas.[7] As FEMA Administrator Craig Fugate stated in testimony before the Senate Indian Affairs Committee:

> The updated policy reiterates the Agency's view of tribal governments as inherently sovereign nations and not political subdivisions of states. To this end, and to the extent permitted by law, FEMA consults with tribal governments and addresses any concerns before taking actions that may affect those nations.
>
> In addition, the new policy expressly states that FEMA will identify and take reasonable, appropriate steps to eliminate or diminish procedural impediments to working directly and effectively with tribal governments. In particular, the policy states that FEMA will review portions of the Robert T. Stafford Disaster Relief and Emergency Assistance Act, and other laws, policies, and administrative rules in emergency management activities to determine how FEMA may work more directly with local tribal communities.[8]

Currently, the Stafford Act includes "an Indian tribe or authorized tribal organization, or Alaska Native village or organization" in the definition of local governments.[9] With this amendment, tribes are also now equivalent to states in their ability to request a major disaster declaration or an emergency declaration from the President.

Tribes had sought this authority for various reasons. While tribes and Native Americans have long received assistance under Stafford Act declarations, working through the state government for all assistance has been viewed as an issue of tribal sovereignty. States might at times be reluctant to request on behalf of a tribe when the damage was localized on tribal property. Other challenges to administering disaster relief involved language barriers and the physical isolation of some tribal lands.[10] Also the tribes wished to have the same ability as states to help manage the response and recovery from a disaster. All of these factors created challenges for emergency management following disaster events in tribal areas.

Under this amendment, the "Chief Executive of an Indian Tribal Government" is able to submit a request for a declaration by the President. In addition, the "Savings provision" of this section ensures that a tribal government is not prohibited from receiving assistance under a declaration made by the President at the request of the governor, if the President has not made a separate declaration for the tribal government. In effect, a tribal government will retain the ability to be treated as a local government in those situations.

Cost Share Adjustments for Indian Tribal Governments

Section 1110 also includes a cost-share adjustment for tribal governments. Under current law, states must provide up to 25% of certain disaster costs, with some adjustments possible under regulatory guidelines.[11] Section 1110 of P.L. 113-2 adds a new Section 401(c)(1) to the Stafford Act which permits the President to:

> waive or adjust any payment of a non-federal contribution with respect to the assistance if –
>
> (A) the President has the authority to waive or adjust the payment under another provision of this Act; and
>
> (B) the President determines that the waiver or adjustment is necessary and appropriate.

The degree of flexibility that is available for the President to set the cost-share for a tribal government is dependent upon the type of assistance being provided. For example, this new Section 401(c)(1)(A) of the Stafford Act appears to retain the cost-share under Section 404 for the Hazard Mitigation Grant Program which states that the President may contribute "up to 75 percent of the cost of hazard mitigation measures."[12] However, the President may have greater discretion under this provision to waive cost-shares for tribal governments for the Public Assistance (Section 406) program which, by FEMA regulation, adjusts to a 90% federal, 10% state and local cost share.[13] As with Section 206 of the Stafford Act, the President is charged with establishing the criteria for such adjustments.

Changes to the Individual Assistance Program

Individual Assistance Factors

FEMA's past judgments involving the Individual Assistance program and the declaration process have provoked interest in the factors FEMA considers when deciding whether to provide Individual Assistance during disasters and how those factors are interpreted. Since it helps determine the extent of federal supplemental assistance, both Congress and state governments share an interest in the declaration policy. Section 1109 of the Sandy Recovery Improvement Act of 2013 calls for FEMA to review its existing regulations that set forth the factors it considers when deciding whether to provide Individual Assistance during disasters (44 C.F.R. §206.48). This review is required within one year of enactment (which is January 29, 2014).

One significant argument for mandating such a review of Individual Assistance factors is that these factors have not been adjusted since they first appeared in regulation in 1999. FEMA is to undertake this review in cooperation with representatives of state, tribal, and local emergency management agencies. The intent of the section is to speed up the declaration process through this review. The current factors that FEMA considers when considering a declaration or a designation within a declaration are listed in its regulations.[14] No special weight is assigned to any of the six factors as they are listed in the regulations.[15] This provision directs that FEMA give special attention to the second factor listed, trauma, when FEMA is measuring the "severity, magnitude and impact of a disaster." Currently, the regulation defines trauma as follows:

> We consider the degree of trauma to a state and to communities. Some of the conditions that might cause trauma are:

(i) Large numbers of injuries and deaths;
(ii) Large scale disruption of normal community functions and services; and
(iii) Emergency needs such as extended or widespread loss of power or water.[16]

Each of the existing factors considered for an IA declaration can be important in the decision-making process but the reference to the "extended or widespread loss of power" in the trauma factor section speaks to recent experiences in many states and wide-spread frustrations during disaster situations. Those frustrations are exacerbated during long power outages and perceived delays in assessing the impact of a disaster.

Child Care

Section 1108(a) of the Sandy Recovery Improvement Act of 2013 also impacts the Individual Assistance program by amending Section 408 of the Stafford Act. Since Hurricane Katrina, a great deal of attention has been given to the plight of children during natural disaster events. One of the areas under consideration has been how to provide child care assistance during especially trying circumstances. One of the challenges is that many such providers of these services are private entities and the Stafford Act assists only government and non-profit organizations. As the Commission on Children and Disasters noted:

> Expanding federal grant eligibility may be justified because many of the services provided by private entities such as child care facilities and physical and mental health clinics provide a service that fills a gap left by limited local, state, and federal resources.[17]

In approaching how to address the identified needs, Section 1108(a) includes "child care" as an eligible expense under the Other Needs Assistance (ONA) program. ONA is part of Section 408 of the Stafford Act, the Federal Assistance to Individuals and Households Program (IHP).[18] IHP is best known as the authority for FEMA's temporary housing program but ONA is also an important part of the program that attempts to address immediate needs that are not met by insurance or other means.

ONA is emergency assistance that helps a family or individual address immediate needs. The provision adds "child care" to a list of eligible activities that currently includes medical, dental and funeral expenses. ONA can also include replacement of personal property (such as clothing and furniture), transportation costs, and other expenses deemed necessary for recovery.

While the federal government pays all housing costs, expenses under ONA are cost-shared with the state on a 75% federal, 25% state basis. IHP assistance, that is, temporary housing and ONA, is limited to 18 months and total assistance of $31,900 per household.[19]

Federal Assistance to Individuals and Households

Pilot Program Made Permanent

Section 1103 of the Sandy Recovery Improvement Act of 2013 would make permanent a pilot program initiated by PKEMRA (P.L. 109-295).

> The Individuals and Households Pilot Program authorizes the President, through the FEMA Administrator, to increase the use of existing rental housing to provide temporary housing for victims of major disasters. Through the pilot program, which expires December 31, 2008, the Administrator is to provide for the repair and improvement of multi-family rental properties in disaster areas to increase the rental stock available to disaster victims in the immediate area. The FEMA Administrator may enter into lease agreements with the owners of multi-family units to achieve FEMA's housing goals, with specified restrictions.[20]

As noted previously, the Individuals and Households Program (IHP) consists of temporary housing and the Other Needs Assistance (ONA) program. Section 1103 increases FEMA's options for providing temporary housing under IHP.

It could be argued that the authority to take this approach already existed broadly in the Stafford Act in the section specifying "direct assistance" that mentions the authority to provide temporary housing units "by purchase or lease."[21] However, since that authority has only been used previously for the purchase of manufactured housing, the specific authority established by Section 1103 may be helpful to the recovery process in providing additional options for temporary housing. Given the work necessary to refurbish units, this form of housing may also contribute to the local economy in the disaster-affected area.

Lease and Repair of Rental Units for Temporary Housing

Section 1103 also authorizes FEMA to enter into lease agreements with private owners of multi-unit apartment facilities. The agreement would include the government paying for repairs or improvements to these apartment units to make them habitable. In return, the owners would agree to lease the improved apartments to house eligible disaster victims. The costs of the repairs would be deducted from the total amount of the lease agreement.

Pilot Program Experience

There was a limited piloting of this multi-family rental unit project in Texas following Hurricane Ike and in Iowa after flooding. FEMA's report to Congress noted:

> The total estimated cost for the Iowa project is $76,854. The estimated cost of providing seven manufactured homes for an equal period of time is $439,376. The total estimated savings to the federal government is $363,522, or 83% less than the cost of providing manufactured homes. The total estimated cost for the Texas pilot project is $897,358. The estimated cost to provide 32 manufactured homes for an equal period of time is $2,650,624. The total estimated savings to the federal government is $1,753,266, or 66% less than the cost of providing manufactured homes.[22]

Although FEMA reported favorable results from the pilot program, this may be an option of limited utility depending on the area where the disaster occurs and the availability of undamaged or unrented units. On the other hand, this arguably provides FEMA with another tool for helping to house individuals in their home area while also making expenditures in that same area that can contribute to the overall recovery.

Changes to the Public Assistance Program

Public Assistance Alternative Procedures

In brief, the PA Program helps state and local governments, as well as certain eligible private non-profits, in their response and recovery to a disaster by providing financial assistance in the form of grants to clear debris and repair and rebuild facilities. Following past disasters, most notably Hurricane Katrina, the PA program has been criticized for being cumbersome and too inflexible to address the needs of local governments and communities. These criticisms were highlighted in several hearings in the immediate aftermath of Hurricane Sandy.[23]

Section 1102 of the Sandy Recovery Improvement Act of 2013 revises the Stafford Act by creating a new Section 428 that authorizes the Administrator to establish and adopt alternative procedures for administering federal assistance under the Public Assistance program.[24] The alternative procedures apply to all traditional categories of Public Assistance (PA) except for Category B—Emergency Protective Measures.[25]

The stated goals of the new alternative procedure are to reduce the cost of federal government assistance; increase the administrative flexibility of the PA program; expedite the process of providing and using the assistance; and create incentives for applicants to complete projects in a timely and cost-effective manner.[26] The provision does not *require* the Administrator to adopt and develop such procedures[27] nor would the Administrator be obligated to approve projects under the alternative procedures.[28] Any participation in the adopted alternative procedures is voluntary at the request of the applicant.[29]

The law also mandates several requirements for the alternative procedures.[30] These requirements can be split into two elements; those changes that apply to the repair, restoration, and replacement of damaged facilities under Section 406 of the Stafford Act, and those changes that apply to debris removal assistance under Sections 403(a)(3)(A), 407, and 502(a)(5) of the Stafford Act.[31] For the applicability of the overall alternative procedures prospectively and retrospectively to major disasters, see the "Applicability to Prior Declared Disasters" section in this report.

Alternative Procedures for Section 406 of the Stafford Act

The alternative procedures include three major revisions for how FEMA can provide assistance for the repair, restoration, and replacement of damaged facilities[32] (assistance provided through Section 406 of the Stafford Act). The provision will allow for procedures to issue grants to recipients based on estimates of their eligible cost as opposed to reimbursement for eligible actual costs; procedures revising the in-lieu contribution authorized in Section 406(c) of the Stafford Act; and procedures for consolidating project worksheets for public assistance projects. Each revision is examined in full below.

Grants Based on Estimates of Eligible Costs

Under the new alternative procedures, FEMA is authorized to issue public assistance grants to applicants based on estimates of their total public assistance eligibility. Of note, the authority to provide grants based on estimates of eligible assistance already existed in Section 406(e) of the Stafford Act. This authority for issuing grants based on estimates was added to the Stafford Act by Section 205(d) of the Disaster Mitigation Act of 2000 (P.L. 106-390).[33] In that law, the President was directed by Section 406(e)(3) to convene an expert panel on how costs should be estimated by FEMA, and to issue regulations implementing these cost estimation procedures. Although the expert panel convened and issued a report,[34] the regulations implementing the statute have not been issued.[35] In addition, Section 689j of PKEMRA authorized the creation of a Public Assistance Pilot Program[36] which, in part, similarly allowed the President to issue grants based on estimates.

Despite previously having the authority to issue grants based on estimates, in most situations FEMA reimburses public assistance applicants on an "actual cost" basis instead of using estimates. Therefore, FEMA generally provides the full assistance amount only after the eligible work on a public assistance project has been completed.[37] FEMA does frequently use a cost-estimating process (called the Cost Estimating Format by FEMA) for some PA projects to help anticipate the end obligation for the project. This enables FEMA to anticipate future costs and outlays for PA projects. However, as a general practice, FEMA only issues grants based on those estimates for PA projects when:

- A project is estimated to cost less than $67,500, and therefore is eligible for simplified procedures as authorized in Section 422 of the Stafford Act;[38]
- An applicant in a project is deciding to receive an in-lieu contribution through Section 406(c) of the Stafford Act (an "alternate project"[39] in FEMA terminology); and
- Significant improvements are being planned for the facility (an "improved project"[40] in FEMA terminology).[41]

The authority to issue grants based on estimates of eligible costs will be part of the alternative procedures an applicant can elect to use for PA projects.[42] One important policy issue remains unclear in the law, and will be left to FEMA to interpret in their establishment and implementation of the procedures. The Sandy Recovery Improvement Act does not indicate if the grants are available for any amount of assistance (in terms of dollars), or if the grant assistance will be limited at some dollar amount. The PA Pilot Program authorized by PKEMRA had similarly worded language on grants based on estimates. When FEMA interpreted this similar authority, they implemented procedures that only gave estimated assistance on single facility projects up to $500,000.[43] Therefore, in implementation of these alternative procedures, FEMA may only allow grants based on estimates up to a certain dollar amount. Limiting the availability of grants based on estimates to under a set amount might contradict the legislative intent of this provision, but appears to be a conceivable option given the language and past FEMA precedent.

The Sandy Recovery Improvement Act of 2013 also provides guidance on how estimates could be developed by FEMA. As discussed previously, regulations implementing cost estimation procedures under Section 406(e)(3) of the Stafford Act have not been issued. In absence of such regulations, the alternative procedures include basic guidelines for determining the estimated eligible assistance of each project(s). First, FEMA should, at the applicant's request, accept the estimates of professionally licensed engineers, so long as the estimate complies with FEMA regulations, policy, and guidance.[44] No definition or guidance is provided for what constitutes a "professionally licensed engineer," therefore it will be up to the Administrator to determine what qualifications are required to be professionally licensed. Second, in instances where the estimated cost is over $5 million, the applicant is allowed to request an independent expert panel to review and validate the cost estimate.[45] No further guidance is provided on the composition of the independent expert panel, or the authority this panel has to revise or amend estimated costs. Therefore, these factors will be developed in FEMA's implementation of the procedure.

The law also provides guidance on what will occur in instances where the estimated grant amount does not equal the final actual costs of the applicant's project(s).[46] In instances where the amount provided by grant is less than the actual eligible cost, the applicant will have to agree to pay the overages.[47] In instances where the grant is more than the actual eligible cost of the project(s), FEMA has the authority to allow the applicant to use the extra funds for mitigation activities or other activities improving future disaster relief.[48] It remains unclear what will become of the excess funds if they are not used by the applicant for such prescribed activities. Unlike the cost-estimating procedures called for by Section 406(e)(2)(A) and (B) of the Stafford Act,[49] the guidelines laid out in this section do not include mention of a floor or ceiling percentage by which the actual cost of a project could come in under or over the estimate. In other words, there is no margin of error for the estimate to be different from actual cost.[50]

Some may argue that without an allowance for floor and ceiling percentages (the margin of error), the process of arriving at an acceptable cost estimate in these alternative procedures may be contentious, especially since it is to the applicant's benefit to ensure that the cost estimate is equal to or higher than the ultimate actual costs of the project. The applicant may be incentivized to seek a high estimate because if actual costs come in lower than the estimate, they are able to use the additional funds for other purposes and are also able to avoid having to pay for unaccounted costs without federal assistance (as would be the case if the estimate were too low).

In-Lieu Contribution

Under pre-existing authorities provided in Section 406 of the Stafford Act, state and local governments, as well as owner/operators of eligible private non-profits, may elect to receive an "in-lieu contribution" of federal assistance for PA projects. In brief, applicants can request that FEMA provide, by grant, the amount of assistance that would have normally been provided for the repair, restoration, or replacement of the current facility. This in-lieu contribution/grant can be used to repair or build an alternate facility or for mitigation measures.[51] In Section 406, such a decision by an applicant needs to be in the interest of the public welfare as determined by the applicant. However, FEMA is obligated by law to reduce the amount of assistance provided "in lieu" to the applicant. For public facilities, the reduction is 10% of the eligible cost for repairing the existing facility, for private non-profit facilities, the reduction is 25%.[52] FEMA applies the reduction to the estimated eligible cost of repairing the current facility, not the estimated costs of new project or mitigation activities.[53] This penalty on the in-lieu contribution has been considered a deterrent to applicants from recovering the facilities in innovative ways as opposed to rebuilding and repairing the facility back to the way it was prior to the disaster.

In the alternative procedures established by the Sandy Recovery Improvement Act of 2013, the in-lieu contributions will not be reduced by 10% for public facilities or 25% for private non-profit facilities.[54] In addition, the alternative procedures do not mention the Section 406 requirement that in-lieu contribution be in the interest of the public welfare, though such requirements may be added by FEMA in their implementation of this option to the alternative procedures.[55]

Consolidating Projects

Finally, the alternative procedures for repair and reconstruction projects will explicitly allow applicants to consolidate multiple individual facilities into a single project. Generally, this policy has previously been used to ease the administrative management of making repairs to multiple similar function facilities—such as all damaged elementary schools—by allowing them to be considered a single project. Using a grant estimating procedure, this could allow a local government, for instance, to receive a grant based on the estimate of damages to nine elementary schools, and use that money to build seven bigger elementary schools instead. However, the law does not specifically require facilities to be of a similar function, so hypothetically a local government could seek to combine a police station project with an elementary school project. How liberally projects could be consolidated remains unknown since the ability to consolidate projects is limited "to the extent determined appropriate by the Administrator."[56]

Alternative Procedures for Debris Removal Assistance

The new Section 428(e)(2) of the Stafford Act applies to debris removal assistance authorized by Sections 403(a)(3)(A), 407, and 502(a)(5) of the Stafford Act. This new section of the Stafford Act reauthorizes similar authorities to those granted by the PA Pilot Program established by PKEMRA. The alternative procedures authorities on debris removal allow the Administrator to:

- Provide grants on the basis of fixed estimates, in a similar fashion to the estimation procedures discussed previously for repair and reconstruction projects.
- Use a sliding scale for the federal share of debris removal based on the time it takes to finish debris removal that provides a federal larger share of the eligible cost to the applicant the quicker they complete the debris removal project, as a means of encourage faster work. For example, if a project is completed within 60 days, the federal government might pay 80% of the eligible cost, but if it is completed in 30 days, it might pay 90% of the cost.
- Allow applicants to recycle debris and use the proceeds from such recycling without reducing the award amount to encourage communities to reap financial and environmental benefits from recycling debris to the greatest extent possible.
- Reimburse state, tribal, and local governments or owner/operators of private non-profits for the base and overtime wages of employees that are performing or administering debris removal projects.[57] In an immediate final rule issued on November 9, 2012, FEMA implemented a similar reimbursement policy for declarations made

because of Hurricane Sandy.[58] This provision implements that policy in the alternative procedures permanently, for all future disasters (not just Hurricane Sandy).

- Provide financial incentives for governments with a FEMA-approved debris removal plan and one or more pre-qualified debris removal contracts prior to a disaster. In implementing this authority during the PA Pilot Program, FEMA offered an additional 5% federal share to applicants if they had the debris removal plan and two pre-qualified debris removal contracts in place.

During the PA Pilot Program, FEMA implemented some form of each of these authorities except that they chose *not* to implement a sliding scale, as they determined the policy may have been too complex to implement and would not have produced a financial savings.[59] In the findings reported by FEMA at the conclusion of the PA Pilot Program, FEMA generally saw positive results from the recycling of debris, financial incentives for debris planning, and reimbursement of wages policies. The positive results from the financial incentives and reimbursement of wage policies were confirmed in an audit conducted by the Inspector General of the Department of Homeland Security.[60] However, FEMA indicated that too few applicants used the grant estimating procedure to determine what impact, if any, it would have had on the efficacy of the assistance.[61]

Report Requirement

Section 1102 of the Sandy Recovery Improvement Act of 2013 requires the Inspector General of the Department of Homeland Security to assess the effectiveness of the alternative procedures for Section 406 of the Stafford Act.[62]

The report will contain recommendations from the Inspector General about whether the procedures should be altered or continued in future legislation. Of note, this report is not required to assess the portion of the alternative procedures for debris removal assistance authorized by Sections 403(a)(3)(A), 407, and 502(a)(5) of the Stafford Act. The report will be issued in three to five years following the enactment of the alternative procedures and will be provided to the committees of jurisdiction for the Stafford Act. The issuance of the report may provide Congress with direction on how it could further revise the Stafford Act to improve the PA program.

Simplified Procedures

Section 422 of the Stafford Act authorizes FEMA to provide PA grants based on federal estimates of eligible costs, as opposed to reimbursing on eligible actual costs.[63] Providing this assistance via a federal estimate is deemed a "simplified procedure." Current law caps the availability of this simplified procedure to projects less than $67,500 for disasters declared after October 1, 2012. The cap is adjusted annually.[64] Projects under this ceiling are often referred to by FEMA as "small projects."

Section 1107 of the Sandy Recovery Improvement Act of 2013 revises Section 422 by requiring the Administrator to analyze whether it would be appropriate to raise the ceiling on small projects, based on a number of considerations including how the threshold impacts "cost-effectiveness, speed of recovery, capacity of grantees, past performance, and accountability measures."[65] The Administrator's findings will be provided in a report to the appropriate committees of Congress no later than one year after enactment (which is January 29, 2014). Following submission of this report, the President is instructed to direct the Administrator to establish a new ceiling for small project eligibility in the appropriate amount. This new threshold would be adjusted annually by the CPI and reviewed and re-established no later than every three years. Therefore, the analysis conducted by the Administrator could result in a new ceiling for simplified procedures being established within a year of enactment and every three years after.[66]

Essential Assistance

Section 1108(b) of the Sandy Recovery Improvement Act of 2013 revises Section 403 of the Stafford Act. Among other items, Section 403 of the Stafford Act authorizes the President and federal agencies to provide assistance for essential services.[67] As listed in the statute, essential assistance includes activities like debris removal, the provision of temporary facilities for schools and other essential community services, and public warning about future risks and hazards.[68] In addition to performing the essential activities themselves in support of communities, federal agencies are authorized to make contributions to state or local governments, or operators of private nonprofit facilities who are providing such essential assistance at a federal share of no less than 75%.[69]

Generally, FEMA refers to such activities eligible under Section 403 as "emergency work," which includes some forms of "debris removal" activities and all "emergency protective measures." FEMA has issued regulations guiding how the federal government will reimburse state and local governments for these activities.[70] Under regulations and policy directives, FEMA has determined that, in general, only the overtime wages of permanent employees working for the state and local governments are eligible for reimbursement (as in, *not* the base pay and benefits or "straight time" of an employee).[71] In contrast, FEMA has determined that the full cost of contract labor for this work is eligible for reimbursement.

In a revision of this previous FEMA policy, the Sandy Recovery Improvement Act of 2013 further specifies in Section 403 that the President may reimburse the base and overtime pay and benefits of permanent employees of state, tribal, and local governments for emergency protective measures. This assistance is limited to "work [that] is not typically performed by the employees" and to "the type of work [that] may otherwise be carried out by contract or agreement with private organizations, firms, or individuals."[72] Therefore, FEMA is authorized to reimburse both the base and overtime pay and benefits for some, but not all, permanent employees conducting emergency protective measures. This new provision does not impact the treatment of wages for private nonprofits and it would continue to allow the reimbursement of overtime and hazardous duty pay of all state and local permanent employees conducting emergency protective measures, consistent with past FEMA policy. Other sections of the Sandy Recovery Improvement Act of 2013 address the reimbursement of wages for debris removal.[73]

Under pre-existing FEMA procedures, some argued that state and local governments are encouraged not to use their own employees for emergency protective measures because contract work is currently fully eligible for reimbursement, while the base wages of state and local governments' employees are not reimbursable. The revisions made by this law for reimbursement may address these concerns over FEMA procedures. However, the new law authorizing reimbursement of base pay and benefits would only apply to the permanent employees of state and local governments who do not typically perform emergency protective measures.[74] Therefore, following similar logic as before, state and local governments may still be encouraged not to use their employees who typically conduct emergency protective measures in favor of their employees who do not typically do that work, or to continue to contract for emergency protective measures. Therefore, many of the implications of this policy will depend on FEMA's interpretation of which state and local government employees "typically" conduct emergency protective measures and which do not, as well as what work "may otherwise be carried out by contract or agreement."[75]

Unified Federal Review[76]

Activities necessary to recover or rebuild a community in the wake of a disaster may be subject to various local, state, tribal, or federal laws intended to protect cultural, natural, or environmental resources (e.g., the National Historic Preservation Act, Endangered Species Act, or Clean Water Act). Section 1106 of the Sandy Recovery Improvement Act of 2013 revises the Stafford Act to create a new Section 429 and would require the President, in consultation with the Council on Environmental Quality (CEQ) and the Advisory Council on Historic Preservation, to establish a Unified Federal Review process to address potential delays related to federal compliance with requirements applicable to activities associated with disaster recovery projects. This review process is required to be established within 18 months of enactment. Most federal agencies currently have such procedures in place. Also, CEQ has provided guidance to federal agencies regarding procedures to expedite compliance with applicable environmental review requirements of the National Environmental Policy Act.[77]

Changes to the Hazard Mitigation Program

Section 1104 of the Sandy Recovery Improvement Act of 2013 amends Section 404 of the Stafford Act, which authorizes the Hazard Mitigation Grant Program (HMGP).[78]

Streamlined Procedures

This provision seeks to accelerate the implementation of HMGP assistance. The new Section 404(d)(2) of the Stafford Act requires the utilization of streamlined procedures in order to provide assistance more rapidly. The program has traditionally moved at a slower pace than other disaster assistance given its processes and the types of projects it funds. But the slower pace has troubled public officials in affected disaster areas.[79]

One of the streamlining procedures that will be utilized is to treat multiple structures, generally clusters of houses, as a group. This procedure has some precedent since the Section 404 buy-out program has purchased large swaths of housing in repetitively flooded areas.[80] While each transaction would involve a voluntary commitment on the part of a homeowner for their participation, the packaging of parcels together is a common practice in seeking to remove vulnerable properties and prevent repeat federal expenditures to an at-risk area.

However, the new Section 404(d)(2) of the Stafford Act also references the Prototype Programmatic Agreement of FEMA as a way of addressing cost-effectiveness and cost-share issues for the multiple structures that may be a part of an HMGP project. FEMA has submitted a draft of a "Prototype" to the Advisory Commission on Historic Preservation (ACHP) but has not yet reached an agreement on the document in accordance with ACHP regulations.[81] FEMA has entered into programmatic agreements with states using the "Prototype" as a basic framework for organizing its work on historic preservation. But the "Prototype" does not address cost-effectiveness and has no influence on any cost-share provisions within FEMA programs.

The streamlined procedures stated in the new provision also include the analysis of environmental impacts to historic properties, cost-effectiveness and fulfillment of cost-share requirements for hazard mitigation measures. It also emphasizes that "adequate resources" should be devoted to this effort. Since FEMA has flexibility in its administration of the Stafford Act that direction may be useful.

However, the areas addressed in this section may be challenging provisions to accelerate. For example, the cost-effectiveness requirement can require substantial research into local history, replacement costs, and potential savings. Also, the cost-share for mitigation within the Stafford Act is not as flexible as other parts of Stafford. For the mitigation cost-share, the Stafford Act states that: "The President may contribute up to 75 percent of the cost-share of hazard mitigation measures."[82] However, the expectation of accelerating the process is partly explained and addressed by another part of the provision which permits the President to advance "not more than 25 percent of the amount of estimated cost prior to eligible costs being incurred."[83] This would increase the initial costs of the front end of the disaster but could arguably also speed up a mitigation process that has appeared cumbersome during previous disasters.

State Administration of HMGP

This provision provides the Administrator the authority to waive notice and rulemaking procedures to expeditiously implement procedures for states to administer the HMGP program. Since states cost-share the program (providing 25% of program costs), and establish state-wide priorities for HMGP funding, it may be considered appropriate that they have the option of administering the program. In fact, that authority was first placed in the Disaster Mitigation Act of 2000, P.L. 106-390 which established the authority for state management of

HMGP.[84] The previous law also asked the President to establish criteria for that management.

Given this waiver authority in P.L. 113-2, it appears that the criteria have yet to be established.[85] So this provision provides the authority for a pilot program for state management of the HMGP program while the Administrator is establishing the criteria. The Stafford Act already notes the state's involvement in HMGP and provides the first direction as to what state criteria might contain. The state's authority to manage HMGP is noted in in Section 404(c). That authority has the minimum criteria listed such as a state's ability and the states own mitigation plan and its overall commitment to mitigation. The current regulations also provide a foundation for state management since the regulations do have the requirements for a state administrative plan for the HMGP program.[86]

Dispute Resolution Pilot Program

It is common during the administration of disaster assistance for disputes to arise regarding the eligibility of various costs or activities for reimbursement. When there is disagreement on the scope of work and the costs involved, public assistance applicants have the ability to appeal FEMA's decisions.[87] The appeal process has two tiers; the initial appeal goes to the FEMA Regional Administrator where the disaster occurred and the second appeal goes to FEMA headquarters for a decision.[88] Some critics have argued that the appeals process is time-consuming. Some have also suggested that, though timelines exist in the law and regulations,[89] they are not consistently applied. As the DHS Inspector General pointed out:

> FEMA does not consistently adhere to timeliness standards governing the amount of time its officials have to provide determinations on appeals. We reviewed appeals that showed, on average, FEMA rendered a decision after about 7 months for first appeals, and after about 10 additional months for second appeals. In some cases, the appeal process spanned several years. For example, a subgrantee submitted a first appeal on February 27, 2005, and received an unfavorable response from FEMA on October 20, 2006, an elapsed time of about 20 months. On March 9, 2007, the subgrantee submitted a second appeal and received a response from FEMA 17 months later, on August 7, 2008. In this example, the total elapsed time for the appeal process was about 3 years.[90]

In light of these concerns, some have suggested that arbitration, or other forms of alternative dispute resolution (ADR), may be a more efficient alternative to the traditional appeals process.

In 2009, the American Recovery and Reinvestment Act (ARRA) directed FEMA to establish arbitration panels as an alternative to the traditional administrative appeals process to facilitate resolution of claims arising from Hurricanes Katrina and Rita.[91] On August 31, 2009, FEMA promulgated regulations implementing these arbitration panels.[92]

More generally, Section 1105 of the Sandy Recovery Improvement Act of 2013 directs the Administrator of FEMA to establish a pilot program containing procedures under which an applicant may request the use of alternative dispute resolution. FEMA also has a pre-existing ADR program, under which ADR specialists or mediators use a variety of methods to facilitate an agreement between the parties.[93] Section 1105 specifies that the ADR procedures shall include arbitration by an independent review panel (IRP), but does not appear to preclude other modes of ADR.[94] Nevertheless, the majority of the provision deals specifically with the IRP, suggesting a particular focus on arbitration.

Eligibility for ADR

Under the provision, not every dispute will be eligible for resolution using the ADR procedures to be established under the pilot program. Eligible disputes must involve assistance provided under sections 403,[95] 406,[96] and 407[97] of the Stafford Act. Notably, the new Stafford Act Section 428 (discussed previously) is not part of the list of covered sections. However, because Section 428 of the Stafford Act does not itself authorize any assistance, it is likely that projects approved under Section 428 would still be viewed as providing assistance under the authority of Section 403, 406, or 407 and therefore would not be categorically ineligible for the new ADR procedures.

Disputes must also involve a large enough amount of money to qualify for the pilot program. Specifically, the "legitimate amount in dispute" must be at least $1,000,000, adjusted annually to account for inflation.[98] The phrase "legitimate amount in dispute" is not defined, but suggests that an applicant may not simply allege a disputed amount above $1,000,000 in order to be eligible for ADR. It also suggests that the threshold is measured not by the aggregate size of a particular project, but is limited to the "amount in dispute." In contrast, the arbitration panels for Hurricanes Katrina and Rita PA claims had jurisdiction over disputes involving projects that totaled more than $500,000, even if only a portion of that total was actually disputed.[99]

ADR under the pilot program is also limited to situations where the assistance is provided with a non-federal cost share. Frequently, several types of assistance provided through the Stafford Act are provided under a 100% federal share, especially under section 403 for essential assistance (particularly categories A and B of assistance). Therefore, many claims involving federal assistance may not be eligible for the dispute resolution procedures provided under the pilot program. This provision also restricts eligibility to those projects that have already received a decision on first administrative appeal.[100]

Independent Review Panels

Under the pilot program, several requirements are imposed upon the use of an IRP, presumably in the context of arbitration.[101] The decisions of the IRP will be binding upon both FEMA and the applicant. Additionally, an applicant that has requested an IRP must waive any remaining appeals regarding the dispute. Similar waiver requirements were adopted by regulation when implementing the arbitration panels for Hurricanes Katrina and Rita PA claims established under ARRA.

Section 1105 makes the Administrator responsible for designating a sponsor for the IRPs who must be unaffiliated with the dispute. However, it is explicitly permitted for the sponsor to be another federal agency, an administrative law judge, or a former federal employee. For the arbitration panels for Hurricanes Katrina and Rita PA claims, FEMA had previously used judges that sit on the Civilian Board of Contract Appeals, which oversees disputes between federal agencies and government contractors.[102] The size and composition of the panels is not fixed under Section 1105.

When considering any arbitration scheme, an important question is the appropriate level of review the arbiters would undertake when evaluating disputed determinations made by a federal agency. For example, in one of the first decisions published by the Hurricanes Katrina and Rita PA claims, *Bay St. Louis-Waveland School District*,[103] FEMA argued that the panel's review should be limited to examining whether FEMA's initial determination was either arbitrary or capricious under the APA. Under this standard, the arbiters could not "substitute [their] judgment for that of the agency"[104] and would be required to uphold the agency's determination so long as there was no "clear error of judgment."[105]

However, the arbitration panel rejected FEMA's argument and held that its review would be *de novo* and its determination did not need to defer to

FEMA's initial findings. In support of this assertion, the panel in *Bay St. Louis-Waveland School District* noted that:

> FEMA's own implementing regulations contemplate that a record *be created specifically for the arbitration panel* which will enable the panel to resolve disputes related to a public assistance grant. The record consists of materials submitted by all parties to the arbitration as well as any independent material from technical and scientific experts that the panel considers necessary to resolve the dispute.[106]

The panel reasoned that its authority to create an independent record would be superfluous if it were to defer to FEMA's initial findings. Therefore, *de novo* review was warranted. Subsequent opinions from arbitration panels have affirmed this conclusion and panels have undertaken *de novo* evaluations of the credibility of witnesses and determinations of feasibility that had previously been made by FEMA officials.

However, such *de novo* review does not appear to be permissible for IRPs established under the new pilot program. Instead, Section 1105 specifically provides that the IRP is limited to consideration of the record at the time of FEMA's decision.

Furthermore, a decision of FEMA's may only be set aside if it is found to be arbitrary, capricious, an abuse of discretion, or otherwise not in accordance with law. If FEMA has decided a question of material fact against the claimant during the first appeal, the IRP cannot reverse this finding unless the decision is "clearly erroneous" based on the record available at the time.

Section 1105 also gives IRPs the authority to order a claimant to pay FEMA the costs of arbitration if the request for ADR is found to be frivolous. Amounts received by FEMA under this authority are to be used to credit any appropriations available for eligible assistance, namely the Disaster Relief Fund.

Sunset and Congressional Oversight

Requests for review by an IRP under Section 1105 must be made before the pilot program expires on December 31, 2015. The Comptroller General is also directed to submit a report to the Senate Committee on Homeland Security and Governmental Affairs and the House Committee on Transportation and Infrastructure analyzing the effectiveness of the program. Among other things, the report is specifically directed to discuss:

1. A determination of the available data to complete the report;
2. An assessment of whether the program expedited or delayed the recovery process;
3. An assessment of whether the program increased or decreased costs;
4. An assessment of the procedures and safeguards established by the IRP and the extent to which such procedures and safeguards were followed;
5. A recommendation on whether any part of the program should be made permanent; and
6. Any other recommendations for modifications to the authority under this subsection.

Strategy for Reducing Costs of Future Disasters

Section 1111 of the Sandy Recovery Improvement Act of 2013 requires the Administrator of FEMA to develop a national strategy "for reducing future costs, loss of life, and injuries associated with extreme disaster events in vulnerable areas of the United States." This strategy will include analysis and recommendations on how the resiliency of the nation can be improved to lessen future disaster loss. This strategy is due to Congress 180 days after enactment (July 28, 2013). In many regards, this required strategy is very similar to the National Mitigation Framework currently in development, but overdue for release, under Presidential Policy Directive 8: National Preparedness (PPD-8).[107] This Framework will address the capabilities necessary to reduce loss of life and property by lessening the impact of disasters, as well as identifying the roles and responsibilities of for federal, state, and local governments and the private sector in achieving set goals for mitigation.[108] If the final version of the National Mitigation Framework is released, FEMA may consider this reporting requirement fulfilled and would not release an ancillary national strategy.[109]

Applicability to Prior Declared Disasters

The Sandy Recovery Improvement Act of 2013 contains no generally applicable effective date. In the absence of explicit language stating otherwise, the effective date of an enacted provision is presumed to be the day on which it was enacted, January 29, 2013, in this case. It is in keeping with existing practice that the changes in law effectuated by the Act will apply to disasters declared on or after this date. However, it is less clear whether, and to what extent, these provisions will apply to disasters declared before January 29, 2013, although support can be found in the text and legislative history for applying at least some of these amendments to the declarations arising out of Hurricane Sandy.

Provisions That Explicitly Address Prior Declared Disasters

Two of the Act's provisions, sections 1102 and 1104, contain specific language relating to the applicability of those provisions to prior disasters. Section 1102, providing alternative procedures for public assistance projects, explicitly provides that the alternative procedures for public assistance projects will apply to public assistance projects "for which construction ha[d] not yet begun on the date of enactment." It is not specified how the determination of when construction began would be made, potentially leaving FEMA with some interpretive flexibility.

Section 1104, providing streamlined procedures for hazard mitigation assistance, states that its procedures apply to previously declared disasters for which the period for processing requests for assistance has not ended as of January 29, 2013. FEMA regulations require requests for hazard mitigation assistance to be sent within 12 months of the date of a disaster declaration.[110] Extensions can be granted, in 30 to 90 day increments, up to an additional 180 days.[111]

The Act does not specify if both of these time frames are to be included in the "period for processing requests for assistance." At a minimum, it likely encompasses the initial 12-month period, placing Hurricane Sandy well within the scope of this provision. It could also be argued that any granted extensions would also be included when determining how far back to apply these new procedures.

Provisions That Do Not Explicitly Address Prior Declared Disasters

While interpreting a statute to have retroactive application in the absence of clear legislative intent is disfavored by the courts, this does not necessarily mean that the provisions of the Act which do not explicitly address applicability will have no effect with respect to disasters that have been declared prior to enactment.

The presumption against retroactivity is strongest where application of the new law "would impair rights a party possessed when he acted, increase a party's liability for past conduct, or impose new duties with respect to transactions already completed."[112]

However, where these considerations are not present, the Court has upheld retroactive application of laws and regulations to pre-enactment conduct.[113] It is not clear that the provisions of the Sandy Recovery Improvement Act create the type of retroactive concerns that give rise to the presumption against retroactivity.

For example, Section 1103, which authorizes the Administrator to enter into lease arrangements for the purpose of providing temporary housing, would not appear to impair the rights possessed by any party prior to the date of enactment, nor would it increase a party's liability for past conduct or impose new duties with respect to past transactions. Therefore, it is possible that such leases could be obtained with respect to disasters that had been declared prior to the date of enactment.

Additionally, floor statements made by sponsors of the legislation,[114] along with the fact that the relevant division of the Act was entitled the Sandy Recovery Improvement Act, suggest that the enactments made in that division are intended to apply at least to claims arising out of Hurricane Sandy, if not earlier disasters.

The same could be said of Section 1105 (directing the Administrator to create an alternative dispute resolution program that includes arbitration before an independent review panel), Section 1106 (directing the establishment of a unified federal review for environmental and historic requirements), Section 1107 (directing the Administrator to review and modify, as necessary, the dollar thresholds for projects to qualify for simplified procedures), Section 1108 (authorizing the President to offer additional forms of assistance), and Section 1109 (directing the Administrator to review and revise individual assistance factors).

Section 1110 (giving tribes the ability to request a major disaster or emergency declaration) also does not explicitly address its applicability to prior declared disasters. Nevertheless, it is unlikely to have any such effect as the tribal request provision will only operate when there has not already been a declaration of an emergency or major disaster in the relevant jurisdiction.

CONCLUSION

The Sandy Recovery Improvement Act of 2013 revised numerous disaster assistance programs authorized by the Stafford Act. Many of these changes will require FEMA to develop policies and regulations to implement the law. For example, as discussed above, FEMA will determine key features of the new alternative procedures for the Public Assistance Program mandated by Section 1102 of the law (within certain statutory guidelines). As of March 7, 2013, FEMA has not released any proposed or final rulemakings to implement these changes, though future rulemakings are anticipated (in addition to those mandated by the law). Therefore, Congress may wish to oversee the implementation of these reforms through hearings or other inquiries to ensure that the changes to the disaster assistance programs align with the future interests of Congress.

APPENDIX. COMPARISON OF THE SANDY RECOVERY IMPROVEMENT ACT OF 2013 TO THE DISASTER RECOVERY ACT OF 2012

The Sandy Recovery Improvement Act of 2013, as passed by the 113th Congress (P.L. 113-2), included many similar provisions to the Disaster Recovery Act of 2012, as passed by the Senate in the 112th Congress (H.R. 1). The Disaster Recovery Act of 2012 was part of the Senate-passed Disaster Relief Appropriations Act, 2013 that was not taken up for a vote in the 112th House. However, as evidenced by their legislative similarities, it influenced the eventual passage in the 113th Congress. **Table A-1** provides a comparison of how the legislation can be matched, starting from the Sandy Recovery Improvement Act of 2013. Because of the differing structures in the legislation, many of the similarities are not immediately obvious. There were two policy provisions unique to each bill (one each). These provisions are listed at the end of the table. The unique provision of the Sandy Recovery Improvement Act of 2013 is discussed previously in the full report. An explanation of the unique provision of the Disaster Recovery Act of 2012, which did not become law, Section 609(1), is included after the table in this Appendix.

Section 609(l)—Report on Impacts to Government Budgets and the Community Disaster Loan Program

Section 609(l) of Senate-passed H.R. 1 (112th Congress) would have required the recently established Hurricane Sandy Rebuilding Task Force[115] to submit a report to various committees of the House and Senate.[116] This provision was not included in the Sandy Recovery Improvement Act of 2013. Among other issues, this report would have evaluated the impact of Hurricane Sandy on local and state budgets, especially the impact on the revenue streams of these governments.

The shortage of local revenues, and the resulting limitation on the financial capacity and budget of the local government, has been cited as one of the most significant and consistent hurdles to long-term disaster recovery.[117] The Task Force's report would also have evaluated whether loan assistance is available from private sources to compensate this loss in revenue and impact to their budgets, and whether federal resources and assistance is available to address these budget impacts.

More specifically, the Task Force would have been required to evaluate the ability of FEMA's Community Disaster Loan (CDL) program to "effectively and expeditiously" address these budgetary impacts.

The core purpose of the CDL program, currently codified in Section 417 of the Stafford Act (42 U.S.C. §5184, as amended) and administered by FEMA, is to provide financial assistance to local governments that are having difficulty providing government services because of a loss in tax or other revenue following a disaster.

The program assists local governments by offering federal loans to compensate for this temporary or permanent loss in local revenue. The program is funded through the Disaster Assistance Direct Loan Program account, rather than the Disaster Relief Fund that funds the majority of other Stafford Act programs.

The Task Force was instructed to provide a thorough assessment of the CDL program, to include an analysis of the loan size limitation and FEMA's regulations and procedures governing most aspects of the program.[118] They were also instructed to provide recommendations on how the program could be legislatively amended.

If the provision had become law, the report was required within 90 days of enactment.[119]

Table A-1. Similar Provisions of the Disaster Recovery Acts

Comparison of Sandy Recovery Improvement Act of 2013 to the Disaster Recovery Act of 2012	
Sandy Recovery Improvement Act of 2013	**Similar Provision in the Disaster Recovery Act of 2012**
Section 1102, Public Assistance Alternative Procedures	Section 609(c)
Section 1103, Federal Assistance to Individuals and Households	Section 609(j)
Section 1104, Hazard Mitigation	Section 609(b)
Section 1105, Dispute Resolution Pilot Program	Section 609(g)
Section 1106, Unified Federal Review	Section 609(f)
Section1107, Simplified Procedures	Section 609(d)
Section 1108(a), Essential Assistance	Section 609(i)
Section 1108(b), Essential Assistance	Section 609(e)
Section 1109, Individual Assistance Factors	Section 609(h)
Section 1110, Tribal Requests for a Major Disaster or Emergency Declaration under the Stafford Act	Section 609(k)
Section 1111, Recommendations for Reducing Costs of Future Disasters	N/A
N/A	Section 609(1), Report on Impacts to Government Budgets and the Community Disaster Loan Program

Source: CRS analysis of Division B of P.L. 113-2 (113th Cong.), the Sandy Recovery Improvement Act of 2013, and Section 609 of Senate-passed H.R. 1 (112th Cong.), the Disaster Recovery Act of 2012.

End Notes

[1] The Stafford Act is codified at 42 U.S.C. §5121 et seq.

[2] See a discussion of the purpose of the law in the consideration of the originating bill, H.R. 219, "Sandy Recovery Improvement Act of 2013," House debate, *Congressional Record*, daily edition, January 14, 2013, pp. H65-72.

[3] 6 U.S.C. §701 et seq.

[4] For more on the DRF, see CRS Report R40708, *Disaster Relief Funding and Emergency Supplemental Appropriations*, by Bruce R. Lindsay and Justin Murray.

[5] 42 U.S.C. §§5170 and 5191.

[6] Section 102 (7) defined local governments as including "an Indian tribe or authorized tribal organization, or Alaska Native village or organization." Section 401 of Stafford noted that "all requests shall be made by the Governor of the affected state."

[7] Rob Capriccioso, "FEMA Wants Tribes to be Equal to States," *Indian Country Today Media Network,* December 7, 2011, at http://indiancountrytodaymedianetwork.com/article/fema-wants-tribes-to-be-equal-to-states-66244.

[8] Testimony of FEMA Administrator Craig Fugate, in U.S. Congress, Senate Committee on Indian Affairs, *Facing Floods and Fires—Emergency Preparedness for Natural Disasters in Native Communities,* 112th Congress, 1st. Sess., July 31, 2011.

[9] 42 U.S.C. §5122.

[10] For an example of these difficulties, see DHS/FEMA Region VIII—"A disaster on the reservation," at http://www.clintonlibrary.gov/assets/storage/Research%20-%20Digital%20Library/ClintonAdminHistoryProject/21-30/Box%2025/1229761-fema-regions-3.pdf.

[11] For information regarding current cost-share regulations, see 44 C.F.R. §206.47.

[12] 42 U.S.C. §5170c.

[13] For further discussion on cost-shares see CRS Report R41101, *FEMA Disaster Cost-Shares: Evolution and Analysis*, by Francis X. McCarthy.

[14] 44 CFR §206.48(b).

[15] For further discussion of these factors, see CRS Report RL34146, *FEMA's Disaster Declaration Process: A Primer*, by Francis X. McCarthy.

[16] 44 CFR §206.48(b)(2).

[17] For additional information, see CRS Report R41080, *The National Commission on Children and Disasters: Overview and Issues*, by Natalie Keegan.

[18] 42 U.S.C. §5174.

[19] Department of Homeland Security, Federal Emergency Management Agency, "Notice of Maximum Amount of Assistance Under the Individuals and Households Program," 77 *Federal Register*, pp. 61425-61426, October 9, 2012, at https://Federalregister.gov/a/201.-24675.

[20] CRS Report RL33729, *Federal Emergency Management Policy Changes After Hurricane Katrina: A Summary of Statutory Provisions*, coordinated by Keith Bea.

[21] 42 U.S.C. §5174(1)(B)(i).

[22] U.S. Department of Homeland Security, Federal Emergency Management Agency, *Individuals and Households Pilot Program, Fiscal Year 2009 Report to Congress*, May 19, 2009, p. 4.

[23] Most notably, see the testimony of Mark Riley, the Deputy Director of the Governor's Office of Homeland Security and Emergency Preparedness for the State of Louisiana in U.S. Congress, House Committee on Transportation and Infrastructure, *Review of the Preparedness Response and Recovery from Hurricane*, 112th Cong., 2nd sess., December 4, 2012 (official transcript not yet available).

[24] The Public Assistance program is authorized by Sections 403(a)(3)(A), 406, 407, and 502(a)(5) of the Stafford Act. Respectively these sections of the Stafford Act refer to emergency debris removal; repair, restoration, and replacement of damaged facilities; non-essential debris removal; and debris removal (via an emergency declaration).

[25] The eligibility of costs in Category B assistance would be revised in Section 1108(a) of the Sandy Recovery Improvement Act of 2013. Under FEMA policy guidance for PA, there are seven categories. Under "emergency work" there is Category A—Debris Removal and Category B—Emergency Protective Measures, and under "permanent work" there is Category C—Roads and Bridges, Category D—Water Control Facilities, Category E—Buildings and Equipment, Category F—Utilities, and Category G—Parks, Recreational Facilities, and Other Items. These categories are defined in Federal Emergency Management Agency, *Public Assistance Guide*, FEMA 322, 2007, at http://www.fema.gov/public-assistance-policy-and-guidance.

[26] See 127 Stat. 40 and the new §428(c) of the Stafford Act and its subparts.

[27] See 127 Stat. 40 and the new §428(b) of the Stafford Act, and that it states the Administrator "may" adopt alternative procedures in coordination with state, tribal, and local governments, and owners or operators of private nonprofit facilities. This is distinction is important, as in past legislative changes to the Stafford Act the President and/or the Administrator have not always implemented some of the options afforded to them in law.

[28] See 127 Stat. 39 and the new §428(a) of the Stafford Act, which states that the Administrator "may" approve projects.

[29] See 127 Stat. 40 and the new §428(d) of the Stafford Act. Voluntary participation is subject to "procedures determined by the Administrator." It is unclear if an applicant would be able to choose to participate in the alternative procedures project by project, or if the Administrator may require the applicant to use the procedures for one set of projects or for the whole disaster declaration.

[30] See 127 Stat. 40 and the new §428(e) and its subparts of the Stafford Act.

[31] See 127 Stat. 40-41 and the new §428(e)(1) and §428(e)(2) of the Stafford Act, respectively.

[32] There are two types of facilities eligible for assistance: public facilities, which are owned by state or local governments, and private non-profit facilities that provide critical services. These terms are defined in full in Section 102(9) and (10) of the Stafford Act.

[33] 114 Stat. 1564. Likewise, Section 689j of the Department of Homeland Security Appropriations Act, 2007 (P.L. 109-295) authorized the creation of a Public Assistance Pilot Program (PA Pilot Program) which, in part, also allowed the President to issue grants based on estimates of eligible cost. For more on the PA Pilot program, see Federal Emergency Management Agency, *Public Assistance Pilot Program: Program Guidance*, June 2007, at http://www.fema.gov/pdf/ about/regions/regioniii/papilot.pdf.; and see Federal Emergency Management Agency, *Public Assistance Pilot Program: Fiscal Year 2009 Report to Congress*, May 20, 2009, at http://www.fema.gov/library/viewRecord.do?id= 3683.

[34] Federal Emergency Management Agency, *Public Assistance: Expert Panel on Cost Estimating*, Recommendation Report of Federal Advisory Committee 10733, October 2002, at http://coop.fema.gov/pdf/government/grant/pa/ cefrep.pdf.

[35] Recent proposed bills, such as Section 209 of the FEMA Reauthorization Act of 2012 (H.R. 2903, as passed by the House), have called for the President to issue these regulations within 180 days of enactment.

[36] 120 Stat. 1455. For more on the PA Pilot program, see Federal Emergency Management Agency, *Public Assistance Pilot Program: Program Guidance*, June 2007, at http://www.fema.gov/pdf/about/regions/regioniii/papilot.pdf; and Federal Emergency Management Agency, *Public Assistance Pilot Program: Fiscal Year 2009 Report to Congress*, May 20, 2009, at http://www.fema.gov/library/viewRecord.do?id=3683.

[37] FEMA will, under some circumstances, provide advances on assistance funds (see 44 C.F.R. §13.21). For regulations guiding how FEMA currently provides assistance for Section 406 of the Stafford Act, see 44 C.F.R. §206.200-339.

[38] For more on this authority, see the section "Simplified Procedures" of this report.

[39] For more on what FEMA calls an "alternate project," see an explanation at Federal Emergency Management Agency, *Alternative Project*, at. http://www.fema.gov/public-assistance-project-formulation-cost-estimating/alternate-project.

[40] For more on what FEMA calls an "improved project," see an explanation at Federal Emergency Management Agency, *Improved Project*, at http://www.fema.gov/public-assistance-project-formulation-cost-estimating/improved-project.

[41] For more on how and when FEMA currently uses their Cost-Estimating Format, see an explanation at Federal Emergency Management Agency, *Public Assistance: Cost Estimating Format Standard Operating Procedure*, athttp://www.fema.gov/public-assistance-cost-estimating-format-standard-operating-procedure.

[42] See 127 Stat. 40-41 and the new §§428(e)(1) and (2) of the Stafford Act.

[43] See Federal Emergency Management Agency, *Public Assistance Pilot Program: Fiscal Year 2009 Report to Congress*, May 20, 2009.

[44] See 127 Stat. 41 and the new §428(e)(1)(F) of the Stafford Act.

[45] See 127 Stat. 40 and the new §428(e)(1)(E) of the Stafford Act.

[46] In other words, what will happen if FEMA grants more or less assistance based on an estimate than the applicant was ultimately eligible for once the true costs of a project(s) was revealed after completion—often many years later.

[47] See 127 Stat. 40 and the new §428(e)(1)(A) of the Stafford Act.

[48] See 127 Stat. 40 and the new §428(e)(1)(D)(i) and (ii) of the Stafford Act. It is likely that FEMA will issue guidance on the use of excess funds.

[49] 42 U.S.C. §5172.

[50] In the expert panel recommendations created after DMA 2000, the panel determined that an error range of 10% was appropriate to serve as the "floor" and "ceiling" percentage called for in Stafford Act. This means that had FEMA implemented the regulations required by Section 406(e)(3)(C) of the Stafford Act, there would be some flexibility in how situations where the actual cost of was different than the estimated cost of assistance, as described in Section 406(e)(2) of the Stafford Act.

[51] Section 406(c)(1)(B) of the Stafford Act authorizes applicants to use in-lieu contributions to repair other facilities, build new facilities, or to provide hazard mitigation at other sites. For example, if an elementary school was substantially destroyed after a disaster, the local government may decide that instead of rebuilding that particular school (and having FEMA reimburse them for the federal share of the eligible cost of doing so), the community may be better served by using that money to build a new high school or to better protect a nearby police station (perhaps because of shifting demographic needs in their population).

[52] See §406(c)(1)(A) of the Stafford Act for public facilities, and §406(c)(2)(A) for private non-profit facilities. Federal assistance for public facility alternate projects was raised from 75% of the eligible federal share to 90% of the federal share in Section 609 of the Security and Accountability For Every Port Act of 2006 (SAFE Port Act, P.L. 109-347).

[53] Returning to prior example in footnote 51, this means FEMA provides 90% of what they estimated to be the total eligible federal assistance for repairing the original damaged elementary school, *not* 90% of the new eligible costs of building a different high school or mitigating future damages to the police station.

[54] See 127 Stat. 40 and the new §428(e)(1)(c) of the Stafford Act.

[55] In the PA Pilot Program, FEMA was authorized to implement an in-lieu procedure that would have maintained the 90% assistance level for public facilities but did not mention a public welfare component. However, they did not exercise this authority, so it is unclear what impact, if any, the possible exclusion of the public welfare requirement would have for the program. In current law, the "public welfare" determination is made by the applicant (either a state/local government or an owner/operator of an eligible private non-profit), so FEMA does not have set regulations or policies on how this determination must be made. For more on why FEMA chose not to implement the in-lieu contribution policy in the PA Pilot Project, see Federal Emergency Management Agency, *Public Assistance Pilot Program: Fiscal Year 2009 Report to Congress*, May 20, 2009.

[56] As in, subject to FEMA regulations and guidance. See the new §428(e)(1)(C)) of the Stafford Act.

[57] Reimbursement of wages would be similar to the new authorities provided for emergency protective measures, with the noted exception that for debris removal no distinction is made between the types of government employee, and that the assistance is also available to private non-profit owners. For more on this issue, see the section "Essential Assistance" of this report.

[58] See Department of Homeland Security, "Debris Removal: Eligibility of Force Account Labor Straight-Time Costs under the Public Assistance Program for Hurricane Sandy," 77 *Federal Register* 67285, November 9, 2012.

[59] Federal Emergency Management Agency, *Public Assistance Pilot Program: Fiscal Year 2009 Report to Congress*, May 20, 2009.

[60] Department of Homeland Security, Office of Inspector General, *FEMA's Oversight and Management of Debris Removal Operations*, OIG-11-40, Washington, DC, February 2011, at http://www.oig.dhs.gov/assets/Mgmt/OIG_11-40_Feb11.pdf.

[61] Ibid.

[62] This report requirement is found in the new §428(g) of the Stafford Act. This report requirement was unique to the PA Alternative Procedures provision as passed in P.L. 113-2. It is not found in Section 609(c) of the Disaster Recovery Act of 2012.

[63] 42 U.S.C. §5189. Specifically, eligible costs under Section 403, 406, 407, or 502 of the Stafford Act. For more on how grants are issued by estimate, see the "Alternative Procedures for Section 406 of the Stafford Act" section of the report.

[64] The statute set the ceiling at $35,000, adjusted annually according to the Consumer Price Index for All Urban Customers. This figure is now adjusted to $67,500 for all disasters declared on or after October 1, 2012, see Federal Emergency Management Agency, "Notice of Adjustment of Disaster Grant," 77 *Federal Register* 61423, October 9, 2012.

[65] 127 Stat. 46 and the new §422(b)(1)(A) of the Stafford Act.

[66] Though the law indicates that the President shall determine if an increase in the threshold is appropriate (see the new §422(b)(1)(A) of the Stafford Act), the Administrator is instructed to establish the threshold at "an appropriate amount" in §422(b)(2)(A) of the Stafford Act. Conceivably, this amount may not be higher than the previous threshold, if the President and Administrator have determined within reason that the same or lower threshold is appropriate.

[67] Section 403 authorizes many forms of assistance that federal agencies, upon the direction of the President, may provide to address "the immediate threats to life and property resulting from a major disaster" (42 U.S.C. §5170b). Further, federal agencies are authorized to perform work on public or private property that is "essential to saving lives and protecting and preserving property or public health and safety" (§403(a)(3) of the Stafford Act).

[68] See §403(a)(3) of the Stafford Act for a full list of the activities included specifically in the law.

[69] Section 403(a)(4) of the Stafford Act.

[70] See 44 C.F.R. §206.201(b) for the definition of emergency work, and primarily 44 C.F.R. §206.225 for regulations on emergency work. "Debris removal" is subcategory A and "Emergency protective measures" is subcategory B of the Public Assistance program. For a full list of activities FEMA considers eligible under each category, see Federal Emergency Management Agency, *Public Assistance Guide*, FEMA 322, 2007, at http://www.fema.gov/public-assistance-policy-and-guidance.

[71] See primarily 44 C.F.R. §206.228(a)(2) and Federal Emergency Management Agency, *Labor Costs—Emergency Work*, FEMA Recovery Policy RP9525.7, November 16, 2006, at http://www.fema.gov/pdf/government/grant/pa/9525_7.pdf. An exception to this standard is provided in 44 C.F.R. §206.202(f)(1)(ii), which allows for there imbursement of the base salaries of a host-state's permanently employed staff who are supporting evacuations or shelters.

[72] See 127 Stat. 47 the new §§403(d)(1)(A)(i) and (ii) of the Stafford Act.

[73] As noted previously, "emergency protective measures" is a subcomponent of "emergency work" and does not include debris removal, which is a separate category of the PA program. See discussion of the reimbursement for debris removal in "Alternative Procedures for Debris Removal Assistance."

[74] See the clause "not typically performed by the employees" in the new §403(d)(1)(A)(i) of the Stafford Act.

[75] See this clause in the new §403(d)(1)(A)(ii) of the Stafford Act.

[76] This section was prepared by Linda Luther, Analyst in Environmental Policy, 7-6852. For more information, see CRS Report RL34650, *Implementing the National Environmental*

Policy Act (NEPA) for Disaster Response, Recovery, and Mitigation Projects, by Linda Luther.

77 See May 12, 2012, Memorandum for Heads of Federal Departments and Agencies from Nancy Sutley, CEQ Chair, regarding "Emergencies and the National Environmental Policy Act," at http://ceq.hss.doe.gov/ceq_regulations/Emergencies_and_NEPA_Memorandum_12May2010.pdf.

78 42 U.S.C. §5170c.

79 Bruce Eggler, "New Orleans officials frustrated over slow payments from Hazard Mitigation Grant Program", NOLA.com/The Times Picayune, December 16, 2010, at http://www.nola.com/politics/index.ssf/2010/12/ new_orleans_officials_frustrat.html.

80 *Innovative Floodplain Management, Full Mitigation Best Practice Story, Lenoir County, North Carolina,* athttp://www.fema.gov/mitigationbp/bestPracticeDetail.do?mitssid=290.

81 36 C.F.R. §800.14(b)(4).

82 42 U.S.C. §5170c.

83 See 127 Stat. 43 and the new §404(e) of the Stafford Act.

84 P.L. 106-390, §204.

85 The criteria do not appear in regulation or in policy guidance.

86 44 C.F.R. §206.437.

87 42 U.S.C. §5189a.

88 For several years the applicant had three appeals. This was reduced, administratively, to two appeals in 1997 to speed up the overall PA process.

89 Under FEMA regulations, those requesting an appeal have 60 days after they have received notice of the action. Within 90 days after receiving the appeal the regional director (for the first appeal) or the Associate Director (for the second appeal) will notify the appellant of the decision. FEMA may request additional information regarding the appeal in which the appellant has 90 days to provide such information. 44 C.F.R. §206.206(c).

90 U.S. Department of Homeland Security, Office of Inspector General, *Assessment of FEMA's Public Assistance Program Policies and Procedures,* OIG-10-26. December, 2009, at 5, at http://www.dhs.gov/xoig/assets/mgmtrprts/OIG_10-26_Dec09.pdf.

91 American Recovery and Reinvestment Act of 2009, P.L. 111-5, §601, providing:

Notwithstanding any other provision of law, the President shall establish an arbitration panel under the Federal Emergency Management Agency public assistance program to expedite the recovery efforts from Hurricanes Katrina and Rita within the Gulf Coast Region. The arbitration panel shall have sufficient authority regarding the award or denial of disputed public assistance applications for covered hurricane damage under section 403, 406, or 407 of the Robert T. Stafford Disaster Relief and Emergency Assistance Act (42 U.S.C. 5170b, 5172, or 5173) for a project the total amount of which is more than $500,000.

92 44 C.F.R. §206.209.

93 The agency's ADR website notes that ADR can be effective in preventing and resolving disputes involving external parties in many contexts, including disaster projects, hazard mitigation projects, insurance claims and flood mapping, personal injury and property damage claims, rulemaking, contracting, policy development, negotiations, and litigation. FEMA, *ADR Frequently Asked Questions*, Aug. 11, 2010, at http://www2.fema.gov/help/adr/faq.shtm.

94 For a list of other types of alternative dispute resolution, see Office of Personnel Management, *Alternative Dispute Resolution: A Resource Guide*, at http://www.opm.gov/er/adrguide/section1-a.asp.

95 42 U.S.C. §5170b (authorizing essential assistance to meet immediate threats to life and property).

96 42 U.S.C. §5172 (authorizing assistance to repair, restore, or relocate damaged public or private non-profit facilities).

97 42 U.S.C. §5173 (authorizing assistance for debris clearance).

[98] Inflation would be measured by changes in the Consumer Price Index for all Urban Consumers as published by the Department of Labor.

[99] In the Matter of Forrest County Board of Supervisors, CBCA 1772-FEMA, at 2 (Civilian Bd. Contract App. January 20, 2010) (dispute met cost threshold even though amount in dispute was only $202,443 because total project costs exceeded $500,000).

[100] FEMA regulations provide two levels of administrative appeals regarding assistance determinations. 44 C.F.R. §206.206.

[101] Few requirements beyond the threshold eligibility determination are imposed on ADR methods that do not use an IRP.

[102] Federal Emergency Management Agency, *New Arbitration Panels for FEMA Public Assistance Program Concerning Hurricanes Katrina and Rita*, Aug. 9, 2009, at http://www.fema.gov/news/newsrelease.fema?id=49243.

[103] In the Matter of Bay St. Louis-Waveland School District, CBCA 1739-FEMA (Civilian Bd. Contract App. Dec. 8, 2009).

[104] Citizens to Preserve Overton Park, Inc. v. Volpe, 401 U.S. 402, 416 (1971).

[105] Ibid.

[106] In the Matter of Bay St. Louis-Waveland School District, CBCA 1739-FEMA, at 7 (Civilian Bd. Contract App. Dec. 8, 2009) (emphasis added).

[107] The National Mitigation Framework is one of five National Planning Frameworks called for by PPD-8, the others being for prevention, protection, response, and recovery. The National Disaster Recovery Framework has already been finalized under PPD-8, but the remaining Frameworks are overdue for release/update per the deadline of June 30, 2012, put forth in PPD-8. For more on PPD-8 and its history, see CRS Report R42073, *Presidential Policy Directive 8 and the National Preparedness System: Background and Issues for Congress*, by Jared T. Brown.

[108] For more on the anticipated contents of the National Mitigation Framework, see the working draft Department of Homeland Security, *National Mitigation Framework*, Working Draft, Pre-decisional, May 1, 2012.

[109] Precedent for such is found in the National Disaster Recovery Framework, which the Administration considers fulfilling the requirement for a "National Disaster Recovery Strategy" found in Section 682 of PKEMRA (6 U.S.C. §771).

[110] 44 C.F.R. §206.436(d).

[111] 44 C.F.R. §206.436(e).

[112] Landgraf v. USI Film Products, 511 U.S. 244, 280, 114 S. Ct. 1483, 1505, 128 L. Ed. 2d 229 (1994) ("the court's first task is to determine whether Congress has expressly prescribed the statute's proper reach.... When, however, the statute contains no such express command, the court must determine whether the new statute would have retroactive effect, i.e., whether it would impair rights a party possessed when he acted, increase a party's liability for past conduct, or impose new duties with respect to transactions already completed. If the statute would operate retroactively, our traditional presumption teaches that it does not govern absent clear congressional intent favoring such a result.").

[113] For example, Bradley v. School Bd. of Richmond, 416 U.S. 696 (1974) (upholding application of statute authorizing award of attorneys' fees in civil rights litigation to attorneys' services that were rendered prior to the enactment of the statute) and Thorpe v. Housing Authority of Durham, 393 U.S. 268 (1969) (applying eviction procedures promulgated by Department of Housing and Urban Development to pending eviction that had begun before procedures were promulgated).

[114] Rep. Denham, "Sandy Recovery Improvement Act of 2013," House debate, *Congressional Record*, daily edition, vol. 159 (January 14, 2013), p. H68 ("in order to help those communities impacted by Sandy, the FEMA Administrator has made it very clear that these reforms must be signed into law by March 1.").

[115] Executive Order 13632, "Establishing the Hurricane Sandy Rebuilding Task Force," 77 *Federal Register* 74341, December 14, 2012, at https://www.federalregister.gov/articles/2012/12/14/2012-30310/establishing-the-hurricane-sandy-rebuilding-task-force.

[116] These committees are the Committee on Appropriations and the Committee on Homeland Security and Governmental Affairs of the Senate, and the Committee on Appropriations and the Committee of Transportation and Infrastructure of the House of Representatives.

[117] U.S. Government Accountability Office, *Disaster Recovery: Past Experiences Offer Insights for Recovering from Hurricanes Ike and Gustav and Other Recent Natural Disasters*, GA0-08-1120, September 2008, p. 17, at http://www.gao.gov/products/GAO-08-1120.

[118] The regulations for the CDL program are found in 44 C.F.R. §206.360-367.

[119] In addition to the provision discussed above, Section 607 of Senate-passed H.R. 1 (112th Cong.) would have changed the cancellation procedures for CDLs issued after Hurricane Katrina. For more on both provisions and the CDL program, see CRS Report R42527, *FEMA's Community Disaster Loan Program: History, Analysis, and Issues for Congress*, by Jared T. Brown.

In: Sandy Recovery Improvement Act … ISBN: 978-1-62618-770-2
Editor: Magnus Hassan

Chapter 2

FY2013 SUPPLEMENTAL FUNDING FOR DISASTER RELIEF*

William L. Painter and Jared T. Brown

SUMMARY

On January 29, 2013, the Disaster Relief Appropriations Act, 2013, a $50.7 billion package of disaster assistance largely focused on responding to Hurricane Sandy, was enacted as P.L. 113-2.

In late October 2012, Hurricane Sandy impacted a wide swath of the East Coast of the United States, resulting in more than 120 deaths and the major disaster declarations for 12 states plus the District of Columbia. The Administration submitted a request to Congress on December 7, 2012, for $60.4 billion in supplemental funding and legislative provisions to address both the immediate losses and damages from Hurricane Sandy, as well as to mitigate the damage from future disasters in the impacted region.

On January 15, 2013, the House of Representatives passed H.R. 152, the Disaster Relief Appropriations Act, 2013. This bill included $50.7 billion in disaster assistance. This was the third piece of disaster legislation considered by the House in the 113th Congress. H.R. 41, which passed the House and Senate on January 4, 2013 and was signed into law two days later as P.L. 113-1, provided $9.7 billion in additional borrowing authority for the National

* This is an edited, reformatted and augmented version of the Congressional Research Service Publication, CRS Report for Congress R42869, dated February 19, 2013.

Flood Insurance Program. On January 14, the House passed H.R. 219, legislation making changes to disaster assistance programs. The rule for consideration of H.R. 152 combined the text of H.R. 219 with H.R. 152 upon its engrossment, to send them to the Senate as a single package.

The Senate passed H.R. 152 unchanged on January 28, 2013 by a vote of 62-36, and it was signed into law as P.L. 113-2 the next day.

H.R. 152 was not the initial legislative response to the storm. In the 112th Congress, the Senate passed a separate package of disaster assistance totaling $60.4 billion, as well as several legislative provisions reforming federal disaster programs. While appropriations legislation generally originates in the House of Representatives, the Senate chose to act on the Administration's request first by amending an existing piece of House-passed appropriations legislation—H.R. 1. This passed the Senate December 28, 2012, by a vote of 62-32. The House did not act on the legislation before the end of the 112th Congress.

This report analyzes the Administration's request, the initial Senate position from the 112th Congress, and H.R. 152, the legislative package developed in the House that was ultimately enacted as P.L. 113-2. It includes information on legislative provisions as well as funding levels.

INTRODUCTION

On October 25, 2012, Tropical Storm Sandy strengthened to become Hurricane Sandy. The next day, the Federal Emergency Management Agency (FEMA) elevated its ongoing preparedness efforts, sending Incident Management Assistance Teams to states from North Carolina to Vermont. Public and private sector entities began to ramp up efforts to prepare for the storm, including a wide range of federal entities from the Federal Aviation Administration to the Department of Energy. On October 28 and 29, as the storm neared land, the President signed emergency declarations for eight states, as well as the District of Columbia, making federal resources available to help state and local governments as they prepared and as the storm began to impact coastal communities.[1] Hurricane Sandy made landfall in New Jersey the night of October 29, 2012, as a Category 1 Hurricane, with a field of hurricane-force winds 900 miles across.[2]

The storm was responsible for at least 131 deaths in the United States, and damage estimates are still being made. In early November EQECAT, an economic forecasting firm, estimated economic losses from Sandy as $30 billion to $50 billion.[3] As of January 31, 2013, the President had declared major disasters for 12 states as well as the District of Columbia under the authority of the Robert T. Stafford Disaster Relief and Emergency Assistance Act (the Stafford Act).[4]

Given the scale of the damage, the Administration submitted a request to Congress on December 7, 2012, for $60.41 billion in supplemental funding and legislative provisions to address both the immediate losses and damages from Hurricane Sandy, as well as to mitigate the damage from future disasters in the impacted region.[5]

LEGISLATIVE HISTORY

112th Congress

On December 12, 2012, the Senate Appropriations Committee published a draft amendment to H.R. 1[6] on its website that would have provided $60.41 billion in supplemental appropriations. The amendment also included a variety of authorizing provisions sought by the Administration as well as provisions originating in the Senate to modify disaster assistance processes and functions. On December 17, 2012, this proposal was introduced as S.Amdt. 3338.[7] On December 19, the amendment was withdrawn and S.Amdt. 3395, with the same title and overall cost was offered in its place. The Senate amended the amendment, passed it by voice vote and then passed the underlying legislation (H.R. 1) on December 28, 2012, by a vote of 62-32. The House did not act on the legislation before the end of the 112th Congress.

However, one facet of the Administration's request did become law through the 112th Congress. The Administration had sought a legislative provision to increase the bond limit for the Small Business Administration's Surety Bond Guarantees Revolving Fund. A provision increasing the bond limit to $6.5 million, and up to $10 million if a federal contracting officer certified it was necessary, was included in P.L. 112-239, the National Defense Authorization Act for Fiscal Year 2013.[8]

113th Congress

On January 4, 2013, the House and Senate both passed H.R. 41, legislation providing an additional $9.7 billion in borrowing authority for the

National Flood Insurance Program (NFIP), which had been a part of the Administration's request.[9] The President signed it into law as P.L. 113-1 on January 6, 2013.

H.R. 152, which included another portion of the Administration's supplemental request, was introduced on January 4, 2013, and an amendment was filed that same day that included further portions of the original request. The House Appropriations Committee described H.R. 152 as including $17 billion "to meet immediate and critical needs," and the amendment as including $33 billion "funding for longer-term recovery efforts and infrastructure improvements that will help prevent damage caused by future disasters." On January 7, an amendment in the nature of a substitute to H.R. 152 which contained some minor textual changes, along with a restructured "long-term recovery" amendment, was posted on the House Rules Committee website.[10]

The House took up the legislation on January 15, 2013. The amendment with long-term recovery funding passed with several amendments, and the amended bill passed the House by a vote of 241-180. The rule for consideration of the bill combined H.R. 219, a House-passed package of legislative provisions reforming disaster assistance programs, with the appropriations legislation upon engrossment of H.R. 152, and sent them to the Senate as a single package.

The Senate passed H.R. 152 unchanged on January 28, 2013 by a vote of 62-36, and it was signed into law as P.L. 113-2 the next day.

Analysis of the Administration's Supplemental Request and the Legislative Response

Table 1 below outlines the Administration's request for supplemental funding and mitigation funding in the wake of Hurricane Sandy, and the congressional response to those requests. All figures are in millions of dollars of budget authority.

The Administration's request is redistributed by appropriations subcommittee. There is no distinction made in this table for mitigation funding. A breakdown of the Administration's request that illuminates the Administration's separate request for mitigation funding is included in the **Appendix**.

Table 1. FY2013 Disaster Supplemental Request and Congressional Action

By appropriations subcommittee, amounts in millions of dollars of budget authority

Subcommittee / **Bureau**	**Account/ Program**	**President's Request**	**112th Congress**	**113th Congress**
			Senate-passed H.R. 1	**P.L. 113-2 (H.R.152)**
Agriculture, Rural Development, Food and Drug Administration, and Related Agencies				
DEPARTMENT OF AGRICULTURE				
Farm Service Agency	Emergency Conservation Program	15	25.09	15
Farm Service Agency	Emergency Forest Restoration Program[a]	23	58.855	23
Natural Resources Conservation Service	Emergency Watershed Protection Program[b]	180	125.055	180
Food and Nutrition Service	Commodity Assistance Program	6	15	6
Commerce, Justice, Science, and Related Agencies				
DEPARTMENT OF COMMERCE				
National Oceanographic and Atmospheric Administration	Operations, Research and Facilities	393	373	140
National Oceanographic and Atmospheric Administration	Procurement, Acquisition, and Construction	100	109	186
DEPARTMENT OF JUSTICE				
General Administration	Office of the Inspector General	0.02	0.02	0
Federal Bureau of Investigation	Salaries and Expenses	4	4	10.02
Drug Enforcement Agency	Salaries and Expenses	1	1	1
Bureau of Alcohol Tobacco Firearms and Explosives	Salaries and Expenses	0.23	0.23	0.23
Federal Prison System	Buildings and Facilities	10	10	10

Table 1. (Continued)

Subcommittee / Bureau	Account/ Program	President's Request	112th Congress	113th Congress
			Senate-passed H.R. 1	P.L. 113-2 (H.R.152)
NATIONAL AERONAUTICS AND SPACE ADMINISTRATION				
	Construction and Environmental Compliance and Restoration	4	15	15
LEGAL SERVICES CORPORATION				
	Payment to LSC	1	1	1
Defense				
DEPARTMENT OF DEFENSE				
Operations and Maintenance	Operations and Maintenance, Army	5.37	5.37	5.37
Operations and Maintenance	Operations and Maintenance, Navy	41.2	40.015	40.015
Operations and Maintenance	Operations and Maintenance, Air Force	8.5	8.5	8.5
Operations and Maintenance	Operations and Maintenance, Army National Guard	3.165	3.165	3.165
Operations and Maintenance	Operations and Maintenance, Air National Guard	5.775	5.775	5.775
Procurement	Procurement of Ammunition, Army	1.31	1.31	1.31
Revolving and Management Funds	Defense Working Capital Funds	24.2	24.2	24.2
Energy & Water Development, and Related Agencies				
U.S. ARMY CORPS OF ENGINEERS				
	Investigations	30	50	50
	Construction	3,829	3,461	3,461
	Operations and Maintenance	899	821	821
	Flood Control and Coastal Emergencies	592	1,008	1,008

Subcommittee / Bureau	Account/ Program	President's Request	112th Congress	113th Congress
			Senate-passed H.R. 1	P.L. 113-2 (H.R.152)
	Expenses	0	10	10
Financial Services and General Government				
GENERAL SERVICES ADMINISTRATION				
Real Property Activities	Federal Buildings Fund	7	7	7
SMALL BUSINESS ADMINISTRATION				
	Salaries and Expenses	50	40	20
	Office of the Inspector General	5	5	5
	Disaster Loan Program Account	750	760	779
Homeland Security				
DEPARTMENT OF HOMELAND SECURITY				
Customs and Border Protection	Salaries and Expenses	2.402	1.667	1.667
Immigration and Customs Enforcement	Salaries and Expenses	0.855	0.855	0.855
Coast Guard	Operating Expenses	66.844	[d]	[d]
Coast Guard	Acquisition, Construction and Improvements	207.389	274.233	274.233
Secret Service	Salaries and Expenses	0.3	0.3	0.3
Federal Emergency Management Agency	Disaster Relief Fund	11,500	11,484.735	11,484.735
Federal Emergency Management Agency	Disaster Assistance Direct Loan Program	300	300	300
Science and Technology	RDAO	3.249	3.249	3.249
Domestic Nuclear Detection Office	Systems Acquisition	3.869	3.869	3.869
Office of the Inspector General	(by transfer)	0	3	3

Table 1. (Continued)

Subcommittee / Bureau	Account/ Program	President's Request	112th Congress	113th Congress
			Senate-passed H.R. 1	P.L. 113-2 (H.R.152)
	National Flood Insurance Fund[c]	9,700	9,700	0
	General Provisions for this title	0	13	0
Interior, Environment, and Related Agencies				
DEPARTMENT OF THE INTERIOR				
US Fish and Wildlife Service	Resource Management	400	0	0
US Fish and Wildlife Service	Construction	78	78	68.2
National Park Service	Historic Preservation Fund	0	50	50
National Park Service	Construction	348	348	348
Bureau of Safety and Environmental Enforcement	Oil Spill Research	3	3	3
Departmental Operations	Office of the Secretary	0	150	360
ENVIRONMENTAL PROTECTION AGENCY				
	Environmental Programs and Management	0.725	0.725	0.725
	Hazardous Substance Superfund	2	2	2
	Leaking Underground Storage Tank Trust Fund	5	5	5
	State and Tribal Assistance Grants	610	810	600
DEPARTMENT OF AGRICULTURE (FOREST SERVICE)				
Forest Service	Capital Improvement and Maintenance	4.4	4.4	4.4
SMITHSONIAN INSTITUTION				
	Salaries and Expenses	2	2	2

Subcommittee / Bureau	Account/ Program	President's Request	112th Congress	113th Congress
			Senate-passed H.R. 1	**P.L. 113-2 (H.R.152)**
Labor, Health and Human Services, Education, and Related Agencies				
DEPARTMENT OF LABOR				
Employment and Training Administration	Training and Employment Services	50	50	25
DEPARTMENT OF HEALTH AND HUMAN SERVICES				
Administration for Children and Families	Social Services Block Grant	500	500	500[f]
Administration for Children and Families	Children and Families Services Programs	100	100	100[f]
Departmental Management	Public Health and Social Services Emergency Fund	200	200	195[f]
Office of the Inspector General	(by transfer)	0	0	5[f]
SOCIAL SECURITY ADMINISTRATION				
	Limitation on Administrative Expenses	2	2	2[e]
Military Construction, Veterans Affairs and Related Agencies				
DEPARTMENT OF DEFENSE (MILITARY CONSTRUCTION)				
Military Construction	Military Construction, Army National Guard	24.235	24.2	24.235
DEPARTMENT OF VETERANS AFFAIRS				
Veterans Health Administration	Medical Services	21	21	21
Veterans Health Administration	Medical Facilities	6	6	6
National Cemetery Administration		1.1	1.1	2.1
Departmental Administration	IT Systems	0.531	0.5	0.531
Departmental Administration	Construction, Major Projects	207	207	207

Table 1. (Continued)

Subcommittee / Bureau	Account/ Program	President's Request	112th Congress	113th Congress
			Senate-passed H.R. 1	P.L. 113-2 (H.R.152)
Transportation, Housing and Urban Development, and Related Agencies				
DEPARTMENT OF TRANSPORTATION				
Federal Aviation Administration	Facilities and Equipment	30	30	30
Federal Highway Administration	Emergency Relief Program	308	921	2,022
Federal Railroad Administration	Grants to the National Railroad Passenger Corporation	32	336	118
Federal Transit Administration	Public Transportation Emergency Relief Program	11,700	10,777	10,894
Office of the Inspector General	(by transfer)	0	6	6
DEPARTMENT OF HOUSING AND URBAN DEVELOPMENT				
Community Planning and Development	Community Development Fund	17,000	16,990	15,990
Office of the Inspector General	(by transfer)	0	10	10
	TOTAL	**$60,408.669**	**$60,407.418**	**$50,507.684**

Source: CRS analysis of FY2013 Supplemental Appropriations Request, as transmitted in a letter from Jeffrey D. Zients, Deputy Director for Management, to the Honorable John Boehner, Speaker of the House of Representatives, December 7, 2012; H.R. 1, 112th Congress; H.R. 152, 113th Congress.

Notes:

a. The Administration requested funding for the Commodity Credit Corporation (CCC) to carry out program activities authorized under the Emergency Forest Restoration Program. The Senate amendment does not refer to the CCC as the authorized funding mechanism, but rather appropriates funds directly to the Emergency Forest Restoration Program.

b. This is described as funding for "Watershed and Flood Prevention Operations" in the Administration's request.

c. P.L. 113-1 was signed into law on January 6, 2013, providing the $9,700 million in additional borrowing authority requested for the National Flood Insurance Program.

d. Transfer authority is provided to other Coast Guard accounts from Coast Guard Acquisition, Construction and Improvements.

e. The House derives these funds from unobligated balances, therefore they do not add to the bill's budgetary score, according to CBO.

f. P.L. 113-2 appropriates $800 million to the PHSSEF account, but requires the HHS Secretary to transfer specified portions of these funds as follows: $500 million to the SSBG, $100 million to the Head Start program (within the Children and Families Services Programs account), and at least $5 million to the HHS Office of the Inspector General (OIG). The remaining $195 million remains available to the HHS Secretary for other activities in the PHSSEF account.

Headers in bold italics note the Appropriations subcommittee of jurisdiction, followed by the department or independent agency in bold capitals. Two columns then specify where a given appropriation is going, by bureau, if applicable, then account or program. The Administration's request is next, in millions of dollars of budget authority, followed by the appropriations that would have been provided if Senate-amended H.R. 1 from the 112th Congress had been enacted. This is provided only for historical reference, as the bill expired with the end of the 112th Congress. The last column reflects the amount of funding provided in H.R. 152 as it passed both House and Senate and was ultimately signed into law. Where accounts are funded through transfers, that number is shown in the table and the donor account is reduced accordingly.

After the table is an analysis of this supplemental appropriations bill in the context of the Budget Control Act, and a more detailed discussion of the contents of the request and the positions taken by the House and Senate in response to it.

Disaster Relief and Emergency Funding under the Budget Control Act

The Budget Control Act (BCA)[11] changed the way Congress accounted for federal funding for disaster response and recovery. In previous years, Congress provided funds over and above limits on discretionary appropriations by designating additional appropriations as being for emergency needs. Budget authority provided in this manner did not count against funding limitations on discretionary spending in budget resolutions.

Although the BCA included legislation allowing for emergency appropriations, the new law included provisions that outlined separate treatment for disaster relief,[12] as distinct from emergency funding. Funding designated as disaster relief in future spending bills could be "paid for" by adjusting upward the discretionary spending caps. This allowable adjustment for disaster relief is limited, however, to an amount based on the 10-year rolling average of what has been spent by the federal government on relief efforts for major disasters.[13]

This disaster relief allowable adjustment for FY2013 is $11.8 billion. Under the current continuing resolution, the amount of disaster relief that would be provided under the BCA if the CR extended for the year was $6.4 billion. The Administration proposed using the remainder of the allowable adjustment for disaster relief in its supplemental request, and using an emergency funding designation to ensure the remaining resources provided through the request do not count against the FY2013 budget caps.

The Administration proposed designating all of the supplemental funding it sought as an emergency requirement, with the exception of a portion of the request for the DRF, which would be designated as being for disaster relief under the BCA. The Administration noted in the letter accompanying the request that it was unclear how much of the disaster relief allowable adjustment might be available pending the finalization of general FY2013 appropriations, and that therefore these numbers could require adjustment. Senate-passed H.R. 1 proposed that $5,379 million in DRF funding be designated as being for disaster relief under the BCA, with all but $3,461 million (for Army Corps of Engineers construction activities)[14] of the remaining funding in the bill designated as emergency funding.

P.L. 113-2 contains $41,669 million in emergency funding, $5,379 million for the DRF designated as disaster relief, and $3,461 million for Army Corps of Engineers construction activities that would count against the discretionary budget caps.[15]

P.L. 113-2 Appropriations by Subcommittee

This section of the report is organized by alphabetically by subcommittee of jurisdiction. Except where otherwise noted, all numbers are in budget authority rounded to the nearest million.

Agriculture, Rural Development, Food and Drug Administration, and Related Agencies[16]

Both the President's request and H.R. 152 as enacted (P.L. 113-2) included $224 million for programs under the jurisdiction of the Agriculture Appropriations subcommittee. The Senate bill, H.R. 1 as amended, would also have provided $224 million for the same programs. Three of the four programs that received funding under the President's proposal and P.L. 113-2 are for emergency land assistance and typically only receive funding through supplemental appropriations bills, rather than annual appropriations bills. The fourth is a nutrition assistance program. While the President's request and P.L. 113-2 are similar, they are not identical. The difference between the two is that the President's proposal would have provided $150 million for watershed protection mitigation efforts, while P.L. 113-2 added this $150 million to watershed response and recovery. The Senate bill would have divided the $150 million for mitigation between all four programs proposed under response and recovery.

The Emergency Conservation Program (ECP) and the Emergency Forest Restoration Program (EFRP) are administered by the USDA Farm Service Agency (FSA). ECP assists landowners in restoring the productivity of agricultural land damaged by natural disaster. Participants are paid a percentage of the cost to restore the land to a productive state. EFRP assists private forestland owners with damage caused by a natural disaster on nonindustrial private forest land. Both the President's request and P.L. 113-2 provided $15 million for ECP and $23 million for EFRP; the Senate bill would have provided approximately $25 million and $59 million, respectively. Following Hurricane Sandy, USDA made $15.5 million in previously appropriated ECP funds available to producers in counties that received a major disaster declaration pursuant to the Stafford Act. According to press releases, producers in counties without a declaration were still encouraged to sign up in the event that future funds were made available (further discussed below). Similarly, USDA announced that no funding is available under EFRP; likewise, producers were encouraged to apply if future funding becomes available.[17]

The Emergency Watershed Protection (EWP) program and the EWP floodplain easement program are administered by USDA's Natural Resources Conservation Service (NRCS) and the U.S. Forest Service (USFS). The EWP program assists sponsors, landowners, and operators in implementing emergency recovery measures for runoff reduction and erosion prevention to relieve imminent hazards to life and property created by a natural disaster. The EWP floodplain easement program is a mitigation program that pays for permanent easements on private land in order to safeguard lives and property from future floods, drought, and the products of erosion. The President's proposal would have provided $30 million for EWP recovery and response and $150 million for EWP floodplain easements for mitigation. P.L. 113-2 did not include funding for EWP floodplain easements and instead added $150 million to the general EWP program. Similarly, Senate-passed H.R. 1 did not include funding for EWP floodplain easements, but rather would have provided the equivalent of the President's proposed $150 million to the other USDA programs proposed for funding response and recovery efforts, including $125 million for general EWP. Following Hurricane Sandy, USDA released $5.3 million in prior appropriated EWP funds to 11 states to respond to imminent hazards to life and property.[18] The EWP floodplain easement program has not received funding since FY2009 and has no current funding available for mitigation.[19]

The emergency agricultural land assistance programs are funded through supplemental appropriations, rather than annual appropriations. As a result, funding for emergency agricultural land assistance varies greatly from year to year. These programs traditionally do not require a federal disaster designation from either the President or a state official. Recent changes in appropriations and budget law, however, have altered how disaster funding for the programs may be used. Funding appropriated in FY2012 was to be used for major disasters declared pursuant to the Stafford Act. This same Stafford Act requirement was present in P.L. 113-2 with the additional requirement that funding may only be used for expenses related to the consequences of Hurricane Sandy. The Senate bill also included the Stafford Act requirement but only to a portion of the appropriation for all three land assistance programs. The Senate bill did not include P.L. 113-2's requirement that funds only be used for Hurricane Sandy expenses.

The President requested and P.L. 113-2 provided $6 million for the Commodity Assistance Program account—specifically for The Emergency Food Assistance Program (TEFAP).[20] The Senate-passed H.R. 1 would have provided $15 million for TEFAP. TEFAP funding provides USDA commodity foods and administrative funding to food banks and other emergency feeding organizations. In their request for $6 million, the Administration reasoned that "this amount is equivalent to one month's worth of TEFAP entitlement commodities in the affected areas." In annual appropriations, TEFAP funds are typically available for one fiscal year, but Senate-passed H.R. 1 would have allowed the funds to be available through the end of FY2014. P.L. 113-2 did not include this extended availability of funding. In addition, P.L. 113-2 granted USDA flexibility to allocate foods and funds for administrative expenses to the Sandy-affected areas beyond the TEFAP authorizing law's parameters. Senate-passed H.R. 1 carried the same provision.

Commerce, Justice, Science, and Related Agencies[21]

The Administration's request included $513.3 million for the accounts that are traditionally funded by the Commerce, Justice, Science, and Related Agencies (CJS) appropriations bill. The Senate-passed H.R. 1 would have provided $513.3 million for these accounts. P.L. 113-2 provided $363.3 million for the CJS accounts. As outlined in **Table 1**, the Administration's request for the CJS agencies included $493.0 million for the National Oceanic and Atmospheric Administration (NOAA), $15.3 million for the Department of Justice (DOJ), $4.0 million for the National Aeronautics and Space Administration (NASA), and $1.0 million for the Legal Services Corporation (LSC). Senate-passed H.R. 1 would have provided $11.0 million less for NOAA and $11.0 million more for NASA than the Administration's request. P.L. 113-2 provided $167.0 million less than the Administration's request for NOAA, $6.0 million more for DOJ, and $11 million more for NASA.

Some of the specific differences between the Administration's request, Senate-passed H.R. 1, and P.L. 113-2 are as follows.

- The Administration requested $4.0 million for NASA's Construction and Environmental Compliance and Protection account. The Senate-passed H.R. 1 would have provided $15.0 million for this account. P.L. 113-2 provided $15.0 million for this account.
- The Administration requested a total of $393.0 million for NOAA's Operations, Research, and Facilities (ORF) account. The Administration's request would have allocated most funding to mitigation projects that would have enhanced resiliency of coastal communities and ecosystems. The Senate bill would have allocated more funding to repairs, replacement, and enhancement of equipment and facilities. P.L. 113-2, like the Senate bill, allocates more funding to repairs, replacement, and enhancement of equipment and facilities.
    - Specifically, the Administration requested $360.0 million under the ORF account to assess risks associated with storms and flooding, provide technical assistance to improve preparedness and resiliency in coastal communities, improve forecast and modeling capabilities to support mitigation efforts, and stabilize and restore ecosystems. The Administration requested $13.0 million under the ORF account to repair or replace damaged weather observation, weather radio, and ocean observing assets and facilities belonging to the National Ocean Service, National Marine Fisheries Service, and National Weather Service. The Administration also requested $20.0 million to evaluate impacts on natural resources, support mapping and charting missions, and conduct marine debris assessments.

- Senate-passed H.R. 1 would have provided $373.0 million for the ORF account, of which $6.2 million was for repairing or replacing ocean observing and coastal monitoring assets damaged by Hurricane Sandy; $10.0 million was for repairing and improving weather forecasting capabilities; $150.0 million was for evaluating, stabilizing, and restoring costal ecosystems damaged by the storm; $56.8 million was for mapping, charting, damage assessment, and marine debris coordination and remediation; and $150.0 million was for necessary expenses related to fishery disasters declared in 2012.[22]
- P.L. 113-2 provided $140.0 million for the ORF account, of which $50.0 million was for mapping, charting, geodesy services and marine debris surveys for coastal states impacted by Hurricane Sandy, $7.0 million was to repair and replace ocean observing and coastal monitoring assets damaged by Hurricane Sandy, $3.0 million was for providing technical assistance to support state assessments of coastal impacts of Hurricane Sandy, $25.0 million was for improving weather forecasting and hurricane intensity forecasting capabilities, $50.0 million was for laboratories and cooperative institutes research activities associated with sustained observations weather research programs, and ocean and coastal research, and $5.0 million was for necessary expenses related to fishery disasters declared in 2012 that were the direct result of Hurricane Sandy.[23]

- The Administration's request for NOAA included $100.0 million under the Procurement, Acquisition and Construction (PAC) account to support state and local acquisition of land to restore and build coastal resiliency in areas where rebuilding physical infrastructure is not feasible or desirable, and on activities that can increase the protective capacity of natural ecosystems. Senate-passed H.R. 1 would have provided $109.0 million for the PAC account, of which $47.0 million was for the Coastal and Estuarine Land Conservation Program to support state and local restoration in areas affected by Hurricane Sandy, $9.0 million was for repairing NOAA facilities damaged by the storm, $44.5 million was for repairs and upgrades to NOAA hurricane reconnaissance aircraft, and $8.5 million was for improvements to weather forecasting equipment and supercomputer infrastructure. P.L. 113-2 provided $186.0 million for the PAC account, of which $9.0 million was to repair NOAA facilities damaged in the storm, $44.5 million was for repairs and upgrades to NOAA hurricane reconnaissance aircraft, $8.5 million was for improvements to weather forecasting equipment and supercomputer infrastructure, $13.0 million was to accelerate the National Weather Service ground readiness project, and $111.0 million was for a weather satellite data mitigation gap reserve fund.

Defense

The Administration sought $90 million for the Department of Defense in accounts managed by the Defense Appropriations subcommittees in its request for FY2013 supplemental appropriations for repair and replacement of damaged equipment and facilities.

Both Senate-passed H.R. 1 and P.L. 113-2 provided $88 million for the Department of Defense, following the same structure. The only difference between the bills and the request was a slightly more than $1 million reduction in both bills in the $41 million request for Navy Operations and Maintenance funding.

Energy and Water Development, and Related Agencies[24]

The President's request, Senate-passed H.R. 1 in the 112th Congress and H.R. 152 as enacted (P.L. 113-2) all included $5.35 billion in supplemental funds for the U.S. Army Corps of Engineers (Corps) Civil Works program. The Corps receives annual appropriations through the Energy & Water Development Appropriations bill. Major differences between the bills and the President's request are summarized below.[25]

While the three proposals shared the same total level of Corps funding, they differed in distribution of funds across Corps accounts, eligible uses, and availability of funds.[26] The Senate bill and P.L. 113-2 as enacted both designated Corps funding as an "emergency requirement," with the exception of the Corps Construction Account funding.[27] Thus, while the bills' funding for the Corps Construction Account counted against discretionary budget caps, their funding for other Corps accounts did not count against the caps.

For the Investigations account, the President requested $30 million, while the Senate-passed H.R. 1 and P.L. 113-2 both provided $50 million. P.L. 113-2 set aside $29.5 million of these funds for ongoing storm damage reduction studies in Hurricane Sandy-impacted areas of the Corps North Atlantic Division (which spans the Atlantic coast from Maine to Virginia). Senate-passed H.R. 1 would have made $34.5 million available for a similar study, and expanded the study area to include Gulf Coast areas in the Mississippi Valley Division impacted by Hurricane Isaac (principally Mississippi and Louisiana). Senate-passed H.R. 1 also would have provided $15 million for an interagency planning process with federal and nonfederal officials that would have developed plans to address coastal flooding risks and include innovative approaches to long-term stability. P.L. 113-2 provided the Corps $20 million to conduct a comprehensive coastal flood risk study of the Hurricane Sandy-impacted areas of the Corps North Atlantic Division.

For the Construction Account, the Administration requested $3.83 billion, including $9 million for repair of existing Corps construction projects and $3.82 billion in "mitigation" funding for projects to reduce damages from future storms.[28] The Administration proposed allowing the Corps to transfer the funds to other agencies, states, or local governments to implement elements of plans that would have resulted from the studies funded in the Investigation account. Senate-passed H.R. 1 and P.L. 113-2 both agreed with the Administration's request for $9 million for repair of existing projects, but included $3.46 billion for all other construction needs, approximately $360 million less than the Administration's request. The two bills differed in their direction regarding the use of the funding. P.L. 113-2 designated the overall funding allocation for rehabilitation, repair, and construction of Corps projects, while Senate-passed H.R. 1 would have provided the funding for these same efforts as they relate to the "consequences of natural disasters." It would have also allowed for the transfer of up to $499 million in funds to other Corps accounts "to address damages from previous natural disasters, following normal policies and cost sharing."[29] P.L. 113-2 included no such provision.

Both Senate-passed H.R. 1 and P.L. 113-2 designated $2.90 billion of the $3.83 billion for specific construction purposes. The enacted bill set the funding aside for projects that reduce future flood risk and support long-term sustainability in coastal areas of the North Atlantic Division affected by Sandy, while under Senate-passed H.R. 1 funding would have also been available for projects in Gulf Coast areas of the Mississippi Valley Division affected by Hurricane Isaac. The enacted bill provided that any project "under study" by the Corps in the North Atlantic Division for reducing flooding and storm damage in areas affected by Sandy that the Secretary determines is "technically feasible, economically justified, and environmentally feasible," is eligible for funding, provided House and Senate appropriations committees approve such a recommendation. Eligibility for the construction funding in Senate-passed H.R. 1 would have been based on the study demonstrating "that the project will cost-effectively reduce those risks and is environmentally acceptable and technically feasible."

The three proposals also differed in their approach to construction cost sharing. The construction costs of Corps projects for flood control and coastal storm damage reduction generally are shared 65% federal, 35% nonfederal (33 U.S.C. 2213), with the nonfederal entity receiving credit toward its share for the provision of lands, easements, rights-of-way, relocations, and disposal areas (known collectively as LEERDs).[30] Senate-passed H.R. 1 proposed to alter this practice, and instead required that nonfederal sponsors provide 10% of project costs, plus the LEERD costs.[31] P.L. 113-2 included a waiver for ongoing construction activities to be undertaken at 100% federal expense. This waiver applied only to ongoing construction activities funded by the bill, not for other construction projects.[32] Both bills allowed nonfederal costs be repaid over a 30-year period. Both bills waived a requirement for congressional approval for projects that exceed 120% of their authorization of appropriations under §902 of the Water Resources Development Act (WRDA) of 1986 (33 U.S.C. 2280).

Other differences between the three proposals included differences in the Corps Operation and Maintenance (O&M) and the Flood Control and Coastal Emergencies (FCCE) accounts. While the President had requested $899 million for the O&M account, both P.L. 113-2 and Senate-passed H.R. 1 provided $821 million. This account includes expenses for dredging of navigation channels and project repair. P.L. 113-2 limited availability for these funds to expenses related to the consequences of Hurricane Sandy, while O&M funding in Senate-passed H.R. 1 would have been available nationally.

Both bills provided $1.01 billion for the FCCE account, or $409 million more than the Administration requested.[33] While P.L. 113-2 limited these funds to expenses related to Hurricane Sandy, the FCCE amounts in Senate-passed H.R. 1 would have been for "flood, hurricane, or other natural disasters," with $430 million of that amount specified to restore projects impacted by Hurricane Sandy to their design profiles. Therefore, under Senate-passed H.R. 1, remaining FCCE funds would have been available to support Corps emergency expenditures nationwide, including emergency operations preparations for future events. P.L. 113-2 also set aside $430 million to restore projects impacted by Hurricane Sandy to their "design profiles," but made these funds contingent on completion of one of the major studies required pursuant to language in the Investigations Account.[34] Both bills also waived

FCCE project cost limits under §902 of WRDA 1986, similar to the proposed provisions for the Construction Account.

Finally, both P.L. 113-2 and Senate-passed H.R. 1 provided $10 million for the Corps and Assistant Secretary of the Army (Civil Works) expenses for oversight of emergency response and recovery activities. The Assistant Secretary is to use these funds to facilitate monthly reporting to the House and Senate Appropriations Committees on the allocations and obligations of all the aforementioned Corps funding, beginning 60 days after enactment. The Administration's request included no such funding or reporting requirement.

Financial Services and General Government[35]

One consequence of Hurricane Sandy is that properties under the control of the General Services Administration (GSA) may have been damaged or deemed uninhabitable until repairs are made. The President requested $7 million to be deposited in the Federal Buildings Fund (FBF) at GSA for the repair and alteration of GSA properties damaged by Sandy. Senate-passed H.R. 1 would have provided the amount the President requested. P.L. 113-2 provides $7 million to GSA for repairs to properties damaged by Sandy and other real property activities.

The provisions for the Small Business Administration (SBA) in P.L. 113-2 provided $804 million in budget authority. Senate-passed H.R. 1 would have provided $805 million in budget authority, along with legislative language sought by the Administration. Although P.L. 113-2 contained similar provisions to Senate-passed H.R. 1, there are some slight differences between the two bills and the Administration's request. These differences are discussed below and include:

- P.L. 113-2 provided $20 million for salaries and expenses as well as a provision for grants for cooperative agreements with organizations (such as Small Business Development Centers and Women's Business Centers) to provide technical assistance related to disaster recovery, response, and long-term resiliency to small businesses that are recovering from Hurricane Sandy. However, P.L. 113-2 did not specify—as Senate-passed H.R. 1 did— how the funds should be disbursed between salaries and expenses and grants for cooperative agreements.[36]

- With respect to grants for cooperative agreements and technical assistance, P.L. 113-2 retained the provision to waive matching requirements that was proposed in Senate-passed H.R. 1. The designated recipients of the cooperative agreements and grants differed between the two bills. H.R. 1 would have explicitly directed the grants and cooperative agreements for only current recipients of grants and cooperative agreements. P.L. 113-2, on the other hand, directed the grants and cooperative agreements for small businesses that are recovering from Hurricane Sandy. Both P.L. 113-2 and H.R. 1 contained provisions to expedite the delivery of assistance. H.R. 1 would have expedited the delivery of assistance by using a process that relied, to the maximum extent practicable, upon previously submitted documentation. P.L. 113-2 did not mention the use of previously submitted documents as a method for expediting assistance.
- P.L. 113-2 provided $5 million—the same amount proposed in Senate-passed H.R. 1—to the SBA's Office of Inspector General.
- P.L. 113-2 provided $520 million for the Disaster Loan Program Account for the cost of direct loans to small businesses. It also provided $260 million for administrative expenses to carry out the direct loan program, of which $250 million was for direct administrative expenses of loan making and servicing (including salaries), and $10 million was for indirect administrative expenses (such as information technology security, staffing, and financial management expenses). Senate-passed H.R. 1 would have provided $500 million for the Disaster Loan Program Account as well as $260 million for direct and indirect administrative expenses of loan making.

The Administration requests for response, recovery, and mitigation funding in the wake of Hurricane Sandy included a provision for surety bond guarantees. This provision was not included in P.L. 113-2 as enacted because a similar provision was included in P.L. 112-239, the National Defense Authorization Act for Fiscal Year 2013.

Senate-passed H.R. 1 would have amended the Small Business Act to prohibit the SBA from requiring small business owners to use their primary residence as collateral for disaster loans of up to $200,000 relating to damage to or destruction of the small business, or for economic injury to the small business if the SBA determined that the small business owner had other assets with a value equal to or greater than the amount of the loan that could be used as collateral for the loan. The President's request did not address the issue concerning the use of collateral for the loans, and the provision was not included in P.L. 113-2.

Homeland Security

The Administration requested $12,085 million for the Department of Homeland Security (DHS), as well as $9,700 million in additional borrowing authority for the National Flood Insurance Fund. In the opening days of the 113th Congress, both the House and Senate passed P.L. 113-1, a separate piece of legislation providing the additional borrowing authority.

P.L. 113-2 included $12,072 million for DHS, with several slight changes in its structure from the Administration's request. P.L. 113-2 provided almost $11,488 million for the Disaster Relief Fund (DRF),[37] approximately $12 million less than the request. P.L. 113-2 also included a transfer of $3 million from the DRF to the Office of the Inspector General for DHS. $5,379 million of the appropriation for the DRF was designated as "disaster relief" under the Budget Control Act, as requested by the Administration. The remainder of the funding for the DRF (and in this section) was designated as an emergency requirement, and therefore none of the funding in this section counts against the discretionary budget caps.

P.L. 113-2 included $0.7 million less for replacement of Customs and Border Protection equipment (down from the $2.4 million request). It included a larger appropriation and transfer authority for the Coast Guard's Acquisition, Construction and Improvements function to meet costs in the Operating Expenses function, rather than providing the $67 million requested by the President as a separate appropriation.

Senate-passed H.R. 1 had included the same funding levels for these accounts.

The Administration requested $300 million in subsidy loan authority for the Disaster Assistance Direct Loan Program account, which funds the Community Disaster Loan (CDL) program. The CDL program provides loan assistance to local governments in declared disaster areas to help them overcome a loss in revenues. In Senate-passed H.R. 1, $300 million would have been appropriated to the account to subsidize no more than $400 million in direct loan obligations. The Senate also directed that $4 million of the amount can be used for administration of the program.

P.L. 113-2 included these amounts, as well as an additional provision (Section 401) that repurposed approximately $146 million in unused subsidy loan authority for CDLs in the wake of Hurricane Katrina provided in the Community Disaster Loan Act of 2005 (P.L. 109-88). As the eligibility of local governments to get loans under this act had expired, the unobligated subsidy loan authority would have remained unused without this provision, which allows it to be used for CDLs sought pursuant to a major disaster declaration for Hurricane Sandy. Senate-passed H.R. 1 did not include this provision.[38]

Homeland Security Legislative Provisions

Senate-passed H.R. 1 included a number of legislative provisions in its section on homeland security, some of which had been requested by the Administration. One of these—$9,700 million in additional borrowing authority for the National Flood Insurance Program—was enacted separately. The Senate also included a number of general provisions in Senate-passed H.R. 1 that would have amended programs funded through the DRF. The House passed many of these provisions in H.R. 219, which passed the House on January 14, 2013, and was appended to House-passed H.R. 152 at engrossment as directed by the rule governing floor consideration of the supplemental appropriations bill. These provisions were ultimately enacted as part of P.L. 113-2. Several other provisions from Senate-passed H.R. 1 were not taken up by the House as part of their legislative response to Hurricane Sandy.

NFIP Borrowing Authority[39]

In an attempt to protect the financial integrity of the National Flood Insurance Program (NFIP), and ensure that the FEMA has the financial resources to cover its existing commitments following the devastation caused by Hurricane Sandy, both the President's request and Senate-passed H.R. 1 as amended would have provided for an increase of an additional $9.7 billion in borrowing authority for the NFIP, which is now capped at $20.725 billion.

On January 4, both the House and Senate passed H.R. 41, a separate piece of legislation providing this $9.7 billion in additional borrowing authority. This legislation was signed by the President on January 6, 2013 as P.L. 113-1, and no further borrowing authority for the NFIP is included in P.L. 113-2.

As background, in the aftermath of Hurricane Katrina in 2005, Congress passed and the President signed into law legislation to increase the NFIP's borrowing authority to allow the agency to continue to pay flood insurance claims: first to $3.5 billion on September 20, 2005;[40] to $18.5 billion on November 21, 2005;[41] and finally to $20.725 billion on March 23, 2006.[42] The NFIP is currently about $18 billion in debt largely as a result of the claims from Hurricane Katrina.[43]

By law, the NFIP does not operate under the traditional definition of insurance solvency—that is, it has not been capitalized, rates are set at levels that make the program self-supporting for the historic average loss year, losses and operating expenses are paid out of policyholder premiums, and the program does not generate sufficient premium income to cover flood insurance claims and expenses and build a reserve fund for future catastrophic loss years.[44] Consequently, while the program typically generates a surplus in less-than-average-loss years, when faced with insufficient funds to pay claims and expenses in catastrophic loss years, such as occurred in the aftermath of Hurricanes Katrina, Rita, and Wilma in 2005, Midwest floods of 2008, Hurricane Irene and Tropical Storm Lee in 2011, and Hurricane Sandy in 2012, the NFIP must resort to its statutory authority to borrow from the Treasury to pay approved claims.[45]

Disaster Recovery Act of 2012 and the Sandy Recovery Improvement Act of 2013[46]

The final general provision in Senate-passed H.R. 1's homeland security title, Section 609, was entitled the "Disaster Recovery Act of 2012" and included a number of legislative provisions that are beyond the scope of this report to discuss at length. The Disaster Recovery Act of 2012 included a number of provisions that were similar to H.R. 219, the "Sandy Recovery Improvement Act of 2013." These provisions were not necessarily identical, but in general, the provisions of these bills spoke to a number of Stafford Act programs, including:

- the Hazard Mitigation Grant Program[47]—Section 1104 of P.L. 113-2, and Section 609(b) of Senate-passed H.R. 1;
- the Public Assistance program[48]—Sections 1102, 1106, 1107, and 1108(b) of P.L. 113-2, and Section 609(c), (d), (e), and (f) of Senate-passed H.R. 1;
- the Individual Assistance program[49]—Sections 1103, 1108(a), and 1109 of P.L. 113-2, and Section 609(h), (i), and (j) of Senate-passed H.R. 1; and
- the Community Disaster Loan program[50]—Section 609(l) of Senate-passed H.R. 1.

Section 1105 of P.L. 113-2 directs the Administrator of FEMA to establish procedures under which an applicant for certain assistance under the Stafford Act may request the use of alternative dispute resolution. This is identical to Section 609(g) of Senate-passed H.R. 1.

Section 1110 of P.L. 113-2 allows the President to declare major disasters upon the request of a chief executive of affected Indian tribal government, instead of only at the request of a governor of a state. This mirrors Section 609(k) of Senate-passed H.R. 1.

Section 1111 of P.L. 113-2, which mandates a report to Congress from FEMA on reducing costs of future disasters, did not correspond to a section in the Senate legislation. Section 609(l) (regarding the Community Disaster Loan program) and 609(m) (regarding applicability) of Senate-passed H.R. 1 did not have corresponding sections in P.L. 113-2.

Provisions Unique to Senate-Passed H.R. 1

Senate-passed H.R. 1 included a number of provisions not requested by the Administration that were not included P.L. 113-2. Some of these mirrored proposed legislation in the 112th Congress. These included:

- Section 602—Would have allowed the Administrator of FEMA, in consultation with state, tribal, and local governments, to give greater weight to the effects of a disaster on special populations in making determinations on Individual Assistance;[51]
- Section 603—Would have broadened eligibility of certain costs for reimbursement under the Public Assistance program;
- Section 604—Would have accelerated FEMA's cost-share adjustment process for Section 406 and 407 (generally Public Assistance and Debris Removal) of the Stafford Act for Hurricane Sandy;
- Section 605—Would have established a pilot program for the relocation of state facilities from disaster-prone areas;
- Section 606—Would have authorized construction of permanent flood-risk reduction levees on land purchased with Hazard Mitigation Grant Program(HMGP) funds in West North Central States.[52]
- Section 607—Would have directed the FEMA Administrator to re-evaluate Community Disaster Loans (CDLs) issued to local governments in Louisiana and Mississippi following Hurricane Katrina;
- Section 608—Would have allowed Louisiana communities to request DHS Inspector General audits of post-Gustav debris removal projects.

Interior, Environment, and Related Agencies[53]

P.L. 113-2 contained $1.44 billion for accounts within agencies typically funded by the Interior, Environment, and Related Agencies Appropriations bill. Both the President's request and Senate-passed H.R. 1 (from the 112th Congress) had included slightly more—$1.45 billion for these accounts. Of the total in the law, $829.2 million was for specified accounts of agencies within the Department of the Interior (DOI), $0.2 million more than the President's request of $829.0 million and $200.2 million more than the $629.0 million in Senate-passed H.R. 1. The law also contained $607.7 million for certain accounts within the Environmental Protection Agency (EPA), $10.0 million less than the $617.7 million requested and $210.0 million less than the $817.7 million in Senate-passed H.R. 1. Finally, the total in the law, like the President's request and Senate-passed H.R. 1, contained $6.4 million for "related agencies," namely the Forest Service ($4.4 million) and the Smithsonian Institution ($2.0 million).

At the account level, P.L. 113-2 included funding for 11 accounts within seven agencies/offices, as had Senate-passed H.R. 1. The President's request had contained funding for 10 accounts within six agencies/offices, as reflected in **Table 1**. The law, President's request, and Senate-passed H.R. 1 proposed the same level of funding for seven accounts. The differences were as follows. The President sought $1.09 billion for three accounts, including $78 million for Construction within the Fish and Wildlife Service (FWS). The remaining $1.01 billion would have been for "mitigation projects" through the Resource Management account within the FWS ($400.0 million) and the State and Tribal Assistance Grants (STAG) account within EPA ($610.0 million). Together with mitigation funding requested for agencies funded through other appropriations subcommittees, such funding was to be used for projects that would reduce the risk or damage from future disasters, according to the President. Senate-passed H.R. 1 also included $1.09 billion, but for four accounts as follows: FWS Construction ($78.0 million); Historic Preservation Fund, within the National Park Service ($50.0 million); Departmental Operations, within the Office of the Secretary of DOI ($150.0 million); and EPA STAG ($810.0 million). P.L. 113-2 provided slightly less—$1.08 billion—for the same four accounts: FWS Construction ($68.2 million); NPS Historic Preservation Fund ($50.0 million); Departmental Operations ($360.0 million); and EPA STAG ($600.0 million). Neither the law nor Senate-passed H.R. 1 included funding for FWS Resource Management, while the

Administration's request did not include funding for the Historic Preservation Fund or Departmental Operations.

The $600.0 million in P.L. 113-2 for EPA's STAG account are allocated entirely for capitalization grants for the State Revolving Fund (SRF) programs under the Clean Water Act (CWSRF), which received $500.0 million and the Safe Drinking Water Act (DWSRF), which received $100.0 million. Similarly, all of the $810.0 million in Senate-passed H.R. 1 would have been allocated to CWSRF ($700.0 million) and DWSRF ($110.0 million) capitalization grants. The Administration had requested $600.0 million for clean water and drinking water SRF capitalization grants but did not specify an allocation between the two, and $10.0 million for wetlands restoration and other ecosystem enhancements. The Administration stated that legislative language would be needed to target the $600.0 million for the SRF capitalization grants to the affected states for mitigation projects. While no specific language accompanied the Administration's request, P.L. 113-2 contained several terms and conditions for the EPA STAG account.

P.L. 113-2 included a requirement that the states use not less than 20% but not more than 30% of the SRF capitalization grant funds to provide additional subsidization to SRF loan recipients in the form of forgiveness of principal, negative interest loans, or grants, or any combination of these. Senate-passed H.R. 1 had included a requirement that the states must use not less than 50% of the capitalization grant funds for this purpose. Both the law and Senate-passed H.R. 1 also required the SRF funds to be used only for "...eligible projects whose purpose is to reduce flood damage risk and vulnerability or to enhance resiliency to rapid hydrologic change or a natural disaster at treatment works..." or eligible facilities, and other eligible tasks necessary to further such purposes.

Finally, SRF funds in the law are allocated entirely to states in EPA Region 2 for wastewater and drinking water treatment works and facilities impacted by Hurricane Sandy, rather than allocated according to the existing state-by-state allotment formula under the Clean Water Act for the CWSRF or according to needs surveys under the Safe Drinking Act's for the DWSRF. H.R. 1 as passed by the Senate would have allocated CWSRF and DWSRF funds only to states that have received a major disaster declaration for Hurricane Sandy under the Stafford Act. The President's request did not include a similarly explicit statement, but did indicate that funds for SRF grants would be allocated to "affected states."

Senate-passed H.R. 1 also would have waived the normal requirement that states provide a 20% match for the SRF capitalization funds, and would have allowed states to use CWSRF funds for purchase of land and easements necessary for siting of treatment works projects, which is currently not an eligible activity under the Clean Water Act program. Neither of these provisions were included in P.L. 113-2.

Two other accounts that received funding in P.L. 113-2 also contained specific terms and conditions. One account is in the NPS, while the second is in the DOI, Office of the Secretary. First, both the law and H.R. 1 similarly conditioned appropriations for the NPS Historic Preservation Fund, which provides funds for restoring historic districts, sites, buildings, and objects significant in American history and culture. They limited funding to expenses related to the consequences of Hurricane Sandy, including costs to administer the program and costs to states to ensure compliance with Section 106 of the Historic Preservation Act. Section 106 requires federal agencies to consider the effects of projects they carry out, approve, or fund on historic properties. They also stated that grants could be provided only in areas that have a major disaster declaration under the Stafford Act, and that grant recipients would not be required to provide a match for federal funding, which typically is required.

Second, the law and Senate-passed H.R. 1 contained differing provisions for the DOI Office of the Secretary, Departmental Operations, regarding the purposes for which the funds are to be used. The provision in H.R. 1 was broader. Under both the law and H.R. 1, for instance, DOI bureaus and offices are to use funds for necessary expenses related to the consequences of Hurricane Sandy, but under H.R. 1 they also could have been used for other activities related to storms and natural disasters. Under both the law and H.R. 1, funds also are to be used for increasing the capacity of coastal habitat and infrastructure to withstand storms, and for restoring and rebuilding parks, refuges, and other public assets. However, the law specified that these entities are to be national/federal. Senate-passed H.R. 1 would have provided for other uses of the funds, namely protecting natural and cultural values, and assisting state, tribal, and local governments. Other language in the law and H.R. 1 as passed the Senate was similar. In particular, both measures authorized the Secretary of the Interior to transfer the funds to any account in the Department, and required the Secretary to submit to the Appropriations Committees a detailed spending plan for the funds within 60 days of enactment.

Finally, provisions of the law prohibited the use of funds for two different purposes. First, one provision barred the Secretary of the Interior and the Secretary of Agriculture from using funds in the bill to acquire land. Second, another provision prohibited FWS Construction funds from being used to repair seawalls or buildings on islands in the Stewart B. McKinney National Wildlife Refuge.

Labor, Health and Human Services, Education, and Related Agencies

The President's request, Senate-passed H.R. 1, and P.L. 113-2 each called for supplemental funding to be provided to several programs typically funded by the Labor, Health and Human Services (HHS), Education, and Related Agencies' appropriations bill (see **Table 1**). The majority of these funds ($800 million) will go to HHS to support health, mental health, and social services needs in affected states, including costs related to the construction and renovation of damaged health, mental health, biomedical research, child care, and Head Start facilities. However, P.L. 113-2 included a different mechanism for providing these funds than did the President's request. The President proposed for these funds to be appropriated directly to three separate accounts, while P.L. 113-2 appropriated the entire $800 million to one account and required that some of these funds be transferred elsewhere. In addition to funding for HHS, the President's request, Senate-passed H.R. 1, and P.L. 113-2 each called for funds (of differing amounts) to the Department of Labor to support dislocated workers. P.L. 113-2 provides $25 million for employment services and job training for dislocated workers.

Department of Labor[54]

The President requested funds for the Training and Employment Services account within the Employment and Training Administration of the Department of Labor. Specifically, the President requested $50 million for the Workforce Investment Act (WIA) Dislocated Worker (DW) National Reserve to support National Emergency Grants (NEG). Funds from the NEG are used to support employment and training activities, such as job search assistance and job training, for workers dislocated from employment by major economic dislocations, including natural disasters.[55] Senate-passed H.R. 1 differed slightly in two ways from the President's request. First, Senate-passed H.R. 1 would have provided $50 million for the DW National Reserve, but would not have specified that the funds were to be used solely for NEGs, which are funded out of the National Reserve. Second, Senate-passed H.R. 1 would have allowed the Secretary of Labor to transfer up to $3.5 million of the appropriated funds to any other DOL account for other reconstruction and recovery needs related to Hurricane Sandy. P.L. 113-2 included $25 million for the WIA DW National Reserve, did not specify that funds for the DW National Reserve are to be used solely for NEG, and provided that the Secretary of Labor has authority to transfer up to $3.5 million of the appropriated funds to any other DOL account for other reconstruction and recovery needs related to Hurricane Sandy.

Department of Health and Human Services[56]

The President's request, Senate-passed H.R. 1, and P.L. 113-2 each called for $800 million in supplemental disaster funding for HHS programs, for ultimate distribution as follows: $500 million to the Social Services Block Grant (SSBG), $100 million to the Head Start program, and $200 million to the Public Health and Social Services Emergency Fund (PHSSEF) for other HHS programs. However, P.L. 113-2 used a different approach from the other two measures to appropriate these funds. The request and Senate-passed H.R. 1 called for the $800 million to be appropriated directly to the three separate HHS appropriations accounts. By contrast, P.L. 113-2 appropriated the full $800 million directly to one of the accounts (the PHSSEF), requiring the HHS Secretary to transfer portions of these funds to the other programs and activities in amounts largely consistent with the request: $500 million to the SSBG, $100 million to the Head Start program, at least $5 million to the HHS Office of the Inspector General (OIG), and the remaining $195 million to the HHS Secretary for other activities. In addition, in contrast to the request and Senate-passed H.R. 1, P.L. 113-2 made the $800 million available through FY2015.[57]

As noted, the President's request, Senate-passed H.R. 1, and P.L. 113-2 each used the PHSSEF to fund all or part of HHS's response efforts. The PHSSEF is an account managed by the HHS Secretary and used by appropriations committees to fund certain emergency management activities, and to provide one-time funds through emergency supplemental appropriations. It is not authorized in law except through annual appropriations, and has no accompanying regulations or guidance. PHSSEF funds are intended for transfer to HHS institutes, agencies, and offices to carry out activities specified in appropriations laws. The President requested $200 million to the PHSSEF for transfer to support a number of health-related activities throughout HHS, including (1) National Institutes of Health (NIH) grantees for losses to their NIH-funded biomedical research programs; (2) substance abuse and mental health programs; (3) environmental and public health support; and (4) other activities the Secretary deems necessary for response and recovery from storm-related damage. Senate-passed H.R. 1 largely followed this approach. P.L. 113-2 provided $800 million (the entire HHS amount) to the PHSSEF, for transfer as noted above, specifying that of the $200 million for health-related activities, at least $5 million be transferred to the HHS OIG, and the remaining $195 million to other accounts within HHS as determined by the Secretary. The latter amount may be used, in unspecified amounts, for repair and rebuilding of non-federal biomedical research facilities (presumably NIH grantees). PHSSEF funds may not be used for costs that are reimbursable by FEMA or covered by insurance.

The President's request and P.L. 113-2 both included $500 million for the SSBG at the HHS Administration for Children and Families. The SSBG is a flexible source of funding used by states to support a wide variety of social services, ranging from child care to special services for the disabled.[58] The request called for the $500 million to be directly appropriated to the SSBG, while P.L. 113-2 called for these funds to be transferred to the SSBG from an $800 million appropriation to the PHSSEF. Both the request and P.L. 113-2 included special language targeting supplemental SSBG funds to states directly affected by Hurricane Sandy (i.e., waiving the statutory allocation formula) and allowing states to use these funds for the provision of health services (including mental health services), and costs of renovating, repairing, or rebuilding health care facilities, child care facilities, and other social services facilities. In addition, P.L. 113-2 included several other provisions applicable to the SSBG. For instance, the law gives states up to three years to expend these funds, one year longer than the SSBG's standard two-year expenditure period. In addition, as with other funds in the PHSSEF appropriation, P.L. 113-2 allows SSBG funds to be used for obligations incurred prior to the bill's enactment (provided these costs align with purposes specified in the bill) and prohibits these funds from being used for costs that are reimbursable by FEMA or covered by insurance. Senate-passed H.R. 1 included similar (though not always identical) provisions, along with several others not enacted in P.L. 113-2. For instance, Senate-passed H.R. 1 included language allowing states to use up to 10% of their allotments to supplement any other funds available for the costs of compensating employees of health care providers for lost wages as a result of Hurricane Sandy and for supporting the viability of health care providers whose facilities were substantially damaged. Senate-passed H.R. 1 also included language requiring states to follow certain federal regulations on establishing a Notice of Federal Interest in real property, where applicable.

The President's request and P.L. 113-2 both included $100 million for the Head Start program, funded within the Children and Families Services Programs account at the HHS Administration for Children and Families. The Head Start program provides comprehensive early childhood development services to low-income children.[59] The request called for the $100 million to be directly appropriated to Head Start, while P.L. 113-2 called for these funds to be transferred to Head Start out of the $800 million appropriation to the PHSSEF. The request specified that funds would be made available to affected Head Start agencies for costs of renovating, repairing, or rebuilding damaged facilities, as well as for certain services for affected children and families, including costs of transporting children enrolled in now-closed centers to other Head Start programs. P.L. 113-2 did not include any language about damaged Head Start facilities or affected children. However, the overall PHSSEF appropriations language made it clear that these funds are for disaster response and recovery in affected states. To this end, P.L. 113-2 included language explicitly waiving the statutory Head Start allocation formula and clarifying that funds awarded from this supplemental appropriation would not be considered part of a Head Start program's "base grant" in subsequent fiscal years. As with other funds in the PHSSEF appropriation, P.L. 113-2 allowed Head Start funds to be used for obligations incurred prior to the bill's enactment (provided these costs align with purposes specified in the bill) and prohibits these funds from being used for costs that are reimbursable by FEMA or covered by insurance. Senate-passed H.R. 1 included similar (though not always identical) provisions, along with several others not enacted in P.L. 113-2. For instance, Senate-passed H.R. 1 included language specifying that these funds could be used for costs of renovating, repairing, or rebuilding damaged facilities; costs of supportive and mental health services for affected children and families; and costs of technical assistance for affected Head Start centers. Senate-passed H.R. 1 also included a provision (not in P.L. 113-2) that would have waived the program's non-federal matching rules for these funds. According to a press release on the draft Senate bill from the 112th Congress, these funds were expected to support approximately 265 Head Start centers damaged by the hurricane.[60]

Military Construction, Veterans Affairs and Related Agencies

The Administration sought $259 million for military construction activities and the Department of Veterans Affairs (VA) in its request for

FY2013 supplemental appropriations for repair and replacement of damaged equipment and facilities. The request sought $24 million for Army National Guard military construction efforts to repair damaged facilities and utilities at Sea Girt National Guard Training Center, and $236 million for the VA. The largest single project was a $207 million request through the Major Construction account for renovation and repair of the Manhattan VA Medical Center, which experienced severe flooding. This project would ordinarily require congressional authorization to be funded.

Senate-passed H.R. 1 proposed $259 million for these accounts. P.L. 113-2 included $260 million for these accounts, the only difference from the request being an additional $1 million for the VA's National Cemetery Administration to repair storm damage. Both pieces of legislation included language to allow the Army National Guard Military Construction funding and the VA Major Construction funding to be expended on otherwise unauthorized projects.[61]

Transportation, Housing and Urban Development, and Related Agencies[62]

As requested by the President and proposed in Senate-passed H.R. 1, P.L. 113-2 provided over $29 billion for accounts within agencies typically funded by the Transportation, HUD and Related Agencies bill.

Department of Transportation[63]

The President's request included $12.07 billion for accounts within the Department of Transportation (DOT), as did Senate-passed H.R. 1. P.L. 113-2 included $13.07 billion for DOT accounts, $1 billion more than requested. In each case the vast majority of funding was for public transit. While the request and Senate-passed H.R. 1 were similar in total funding, they differed in funding allocation, and P.L. 113-2 differs from both, as shown in **Table 1**. Briefly, P.L. 113-2 provided (a) more funding for Amtrak than the President requested but less than Senate-passed H.R. 1 provided, and (b) more funding for highway repair than either the President requested or the Senate-passed H.R. 1 provided.

For transit assistance, the President requested a total of $11.7 billion, divided between repair and mitigation funding. Both types of funding would go into the recently created Public Transportation Emergency Relief Program (previously, some public transit emergency relief funding could have been provided under the Stafford Act). The President requested $6.2 billion for repairs and $5.5 billion for mitigation; the repair funding request specified that the funding would be provided as a 90% federal match; that funding could also be transferred for use for highway and bridge repairs at the discretion of the Secretary of Transportation; that funding should not supplant private insurance coverage, and that $3 million would go to the Department of Transportation Inspector General for oversight.

Senate-passed H.R. 1 would have provided $10.78 billion, up to $5.38 billion of which could have been transferred by the Secretary of Transportation to be used to mitigate damage to highway and transit facilities from future disasters (which, by inference, assures that at least $5.4 billion is available for repairs). The President's request would have allowed the repair money to also be used for highway infrastructure, with no language concerning mitigation funding, while the Senate bill reversed that, proposing to make the mitigation funding available for transfer to highway projects, with no corresponding language for the repair funding. Also, as with the Amtrak section, Senate-passed H.R. 1 did not include language addressing the issue of not supplanting private insurance.

P.L. 113-2 provided $10.9 billion for the Public Transportation Emergency Relief Program, $2 billion to be made available immediately and the remainder after the Federal Transit Administration publishes interim regulations for the program. Of the total, the Secretary of Transportation may transfer up to $5.383 billion to fund transportation projects to reduce the risk of damage from future disasters in the areas impacted by Hurricane Sandy. The law also is silent about the issue of not supplanting private insurance money.

For repairs to Federal Aviation Administration (FAA) equipment, the President requested $30 million; Senate-passed H.R. 1 included the President's requested funding level, as did P.L. 113-2. This funding will be drawn from the Airport and Airway Trust Fund.

The President requested $308 million for highway repairs, and called for a portion of the funding requested for the Public Transportation Emergency Relief Program to be available for highway repairs in areas affected by Hurricane Sandy at the discretion of the Secretary of Transportation. P.L. 113-2 provided $2.022 billion, over six times the amount requested, and also allows a portion of the funding provided for the Public Transportation Emergency Relief Program to be available for highway (and other types of transportation) disaster mitigation projects. Senate-passed H.R. 1 would have provided $921 million for highway repair, and also allowed for funds to be made available for mitigation projects.

The President requested $32 million for Amtrak, while Senate-passed H.R. 1 would have provided $336 million. Amtrak has estimated that its property damage and business interruption losses will be around $60 million; it has insurance to cover this, with a $10 million deductible, though it may be some time before the insurance claim can be settled. Amtrak has also identified $276 million in mitigation and capacity-expanding activities for rail tunnels into New York City that it would like to undertake. The President's request included language providing that federal funding should not be used to supplant insurance coverage for Amtrak's damages. Senate-passed H.R. 1 would have provided Amtrak the entire sum ($60 million for repairs and $276 million for mitigation and improvements), with no language addressing the insurance issue. P.L. 113-2 provided $32 million for repairs and $86 million for recovery and resiliency projects in the affected area, a total of $118 million, which is more than requested by the President but considerably less than would have been provided by Senate-passed H.R. 1.

Some transit agencies have proposed that, instead of using emergency relief funding to simply restore infrastructure to its pre-disaster condition by replacing equipment that may be antiquated, they take this opportunity (and funding) to install equipment that makes their systems more functional (such as, for example, increasing capacity) as well as more resilient in coping with future emergencies. The new Federal Transit Administration Emergency Relief Program may provide grantees this flexibility, as both Congress and recent administrations have provided similar flexibility for the Federal Highway Administration Emergency Relief Program. Such an approach may raise questions about how the costs of repairs that include system improvements should be allocated between the federal Emergency Relief programs and state and local governments.[64]

Housing and Urban Development[65]

During the last days of the 112th Congress the President requested, and the Senate-passed version of H.R. 1 included, $17 billion in supplemental funding for HUD, all of which would have been appropriated to the Community Development Fund (CDF), the account that funds the Community Development Block Grant (CDBG) program. During the first weeks of the 113th Congress, the House and Senate considered and passed H.R. 152, which included $16 billion for HUD, all allocated to the CDF. The President signed the measure into law as P.L. 113-2 on January 29, 2013.

While P.L. 113-2 and the Administration's request would have set aside CDBG funds for the activities of the Office of the Inspector General (OIG), P.L. 113-2 transferred $10 million for OIG activities, significantly more than the $4 million requested by the Administration. A provision in Senate-passed H.R. 1 also proposed transferring $10 million to fund OIG activities.

P.L. 113-2 did not include a proposed Administration request that would have set aside $2 billion of the total CDBG disaster aid request for mitigation activities. The Senate-passed proposal also included a proposed set-aside of $2 billion for mitigation activities. Consistent with the Administration's request, P.L. 113-2 included a $10 million set aside for salaries and expenses to be used to fund technical assistance and cover the costs incurred by HUD's Office of Community Planning and Development (OCPD) in administering CDBG disaster funds. The Senate-passed bill also recommended transferring $10 million to the OCPD for such activities. P.L. 113-2 allowed HUD to distribute CDBG disaster funds appropriated under the act to the most impacted and distressed areas affected by Hurricane Sandy and other eligible disaster events occurring during calendar years 2011, 2012, and 2013. A similar provision included in Senate-passed H.R. 1 recommended setting aside a specific amount—$500 million—in CDBG disaster funds to address the unmet needs resulting from other (non-Hurricane Sandy) major disasters declared via the Stafford Act that occurred during 2011 or 2012, or for small, economically distressed areas with a disaster declared in 2011 or 2012.

P.L. 113-2 included several terms and conditions that vary from the rules governing the regular CDBG program, but are consistent with language included in Senate-passed H.R. 1. These can be grouped into three broad areas governing the submission and content of disaster plans, allocation and use of funds, and waiver authority. P.L. 113-2:

- directed HUD to promulgate regulations governing the distribution and use of funds within 45 days after passage of this act, including establishing minimum allocations for CDBG grantees;
- required states and local government grantees to submit, and for HUD to approve, disaster plans before CDBG disaster funds may be obligated;
- required that a grantee's disaster plans articulate how proposed activities will support long-term recovery efforts;
- required HUD to certify that state and local government grantee disaster plans include adequate financial controls and procurement

processes that would prevent duplication of benefits; waste, fraud, and abuse; and encourage timely expenditure of funds; and
- directed HUD to allocate one-third of CDBG disaster appropriations provided in the bill to states and local government grantees within 60 days after passage of the bill.

P.L. 113-2 also established conditions and terms for the use of funds, including:

- allowing grantees to use up to 5% of their CDBG disaster grant allocation for administrative expenses;
- prohibiting grantees from contracting out the responsibility for administering the CDBG disaster programs;
- requiring grantees to include performance requirements and penalties when eligible activities are undertaken through the use of contractors or procurement services;
- prohibiting disaster funds from being used for activities that are reimbursable by, or made available by, FEMA or the Army Corps of Engineers;
- requiring grantees to maintain a publicly accessible website identifying how all grant funds are used, including information on contracting and procurement processes; and
- holding harmless a state or community's regular CDBG allocation by ensuring that the amount of such funds awarded to grantees would not be affected by CDBG disaster-assistance allocations.

P.L. 113-2 did not include two provisions that were included in Senate-passed H.R. 1. Specifically, P.L. 113-2 did not include provisions removing the $250,000 ceiling on the amount of CDBG disaster funds that may be used to meet the non-federal cost share of a disaster-related project funded by the Army Corps of Engineers; or limiting disaster recovery assistance to for-profit entities to businesses that meet the Small Business Administration's definition of small business and to public utilities.

Finally, P.L. 113-2 granted HUD broad authority to waive or establish alternative program requirements, except for provisions governing fair labor standards, fair housing, civil rights, and environmental review. However, P.L. 113-2 included two exceptions related to environmental review requirements. Specifically, it allowed CDBG disaster fund grantees who use their funding to meet certain FEMA matching requirements to adopt, without public review, environmental reviews performed by other federal agencies. In cases where a grantee has already performed an environmental review or the activity or project is excluded from an environmental review, P.L. 113-2 explicitly allowed for the expedited release of funds. The law also allowed HUD to reduce, from 70% to 50%, the percentage of funds that must be targeted to activities benefiting low and moderate income (LMI) persons, and allows HUD to reduce the LMI-targeting requirement below 50% only if the grantee can demonstrate a compelling need. Similar provisions were included in Senate-passed H.R. 1.

The President also requested legislative language for one HUD account for which funds were not sought: the tenant-based rental assistance account, which funds the Section 8 Housing Choice Voucher program. Specifically, the President requested that Congress "hold harmless" program administrators (public housing authorities, or PHAs) affected by the disaster when allocating FY2013 voucher renewal and administrative fee funding provided through the regular annual appropriations process. The President requested that disaster-affected PHAs be funded no lower than their FY2012 funding levels. P.L. 113-2, like Senate-passed H.R. 1, included similar language. It provided the Secretary the authority to make adjustments to PHAs' funding levels to "avoid significant adverse funding impacts that would otherwise result from the disaster," at a PHA's request and provision of supporting documentation.

Additional legislative provisions in the THUD section of Senate-passed H.R. 1 would have (1) required DOT and HUD to submit implementation plans within 45 days of enactment and biannually thereafter and (2) required DOT and HUD to notify the House and Senate Committees on Appropriations not less than three full business days before the announcement that a project, state, or locality has been selected to receive a grant award totaling $500,000 or more. P.L. 113-2 included similar provisions, except that the threshold for notifying the congressional appropriations committees about individual grants was raised to $1 million.

General Legislative Provisions

While the Administration indicated a need for legislative language on a number of issues, no draft texts of proposed language was circulated publicly.

There are four general provisions that apply to the appropriations provided in P.L. 113-2. Three of these were generally administrative in nature, as were two of the nine included in Senate-passed H.R. 1—provisions traditionally carried in supplemental appropriations bills with emergency funding.

Internal Control Plans[66]

The President's request included a proposal to require the Office of Management and Budget (OMB) to direct federal agencies to submit internal control plans for the programs receiving supplemental appropriations.[67] The President's request stated that the internal control plans should contain *enhanced* grant management protocols, including quarterly program and financial monitoring, timely submission of single audit reports and grants closeout, and improper payments testing and reporting.

Existing statutory and regulatory provisions, and OMB guidance, already address these grants management practices, so it is unclear what enhancement of grant management protocols might entail.[68] Additionally, the President's request did not include specific provisions for additional resources for federal agencies to implement grants oversight, such as supplemental funds for federal agency inspector general offices or an increase in the allowable management and administration percentage for individual grant programs. The Administration's request also did not identify which programs would be affected by the enhanced protocols.

In the 112th Congress, Senate-passed H.R. 1 included a provision that would have required OMB to issue guidance to federal agencies to develop internal control plans for funds provided by the bill.[69] The bill also included funding for oversight of supplemental funding and certain management and administration activities, however the amounts were provided at the program level and not all programs received additional funding for these activities.[70] P.L. 113-2 requires federal agencies to submit internal control plans to OMB, GAO, agency Inspectors General, and House and Senate Appropriations Committees for all supplemental funding provided therein; and directs GAO to develop the template for the internal control plans.[71]

Improper Payments

The President's request did not specifically address improper payments, but included a provision to ensure the integrity of federal spending. Both P.L. 113-2 and Senate-passed H.R. 1 included a provision[72] that designated all programs and activities funded through the legislation as "susceptible to significant improper payments" under the provisions of the Improper

Payments Information Act of 2002 (IPIA).[73] This designation requires federal agencies to estimate the annual amount of improper payments made under the program and submit the estimates to Congress annually. Additionally, for programs that have estimated improper payments that exceed $10 million, the federal agency is required to develop a report that identifies the causes and corrective actions the agency will take to reduce the improper payments.[74] Several programs that receive funding under the bill are not currently identified as "susceptible to significant improper payments." This provision, therefore, increases the administrative burden on agencies and grant recipients. No specific appropriations to fund compliance with this provision were requested or included.

Two provisions were added to Senate-passed H.R. 1 through the floor amendment process that sought to prohibit payments from funds provided in the bills. One provision would have prohibited payments to individuals who were deceased at the time funds were made available,[75] and another would have prohibited payments to an individual or entity using funds provided under the bill if the individual or entity had a pending "seriously delinquent tax debt."[76] In regards to the tax provision, it was unclear how agencies would have implemented this provision, as there is some question regarding federal agencies' ability to access IRS tax records to screen disaster recipients prior to providing federal disaster assistance. Neither the President's request nor P.L. 113-2 included these provisions.

Trigger to De-Obligate Unexpended Grant Funding

The President's request recommended the withdrawal of grant funds awarded through certain programs if funds were not expended within 24 months of the award. It was unclear exactly which federal grant programs, and what types of grant awards, were the intended objects of this proposal.

Senate-passed H.R. 1 would have directed agencies to identify (for application of the trigger) grants funded through the legislation where funds should have been expended within the 24-month period following the federal agency obligation of funding. The bill would also have required the Director of OMB to issue guidance establishing the methods federal agencies would use to identify grant awards affected by the trigger. Recipients of identified grants would have had to expend funds in the 24-month period following the award. The federal agency would have had to de-obligate any funds remaining unexpended after the 24-month period. Federal agency heads could have requested a waiver of the 24-month expenditure requirement after consultation with the Director of OMB to discuss exceptional circumstances that might justify an extension. It is unclear whether the Senate provision would have required the director to approve the waiver, and what "consultation" might have entailed. Additionally, in the absence of specific language establishing a time frame for the waiver process, grant recipients could have faced uncertainty about whether they could have continued expending funds once the 24-month period had elapsed. This could have resulted in disaster recovery activities coming to a halt while federal agencies debate approval of the waiver.

P.L. 113-2 contains a provision requiring grant recipients to expend funds within the 24-month period following the federal agency obligation of funds for the grant award unless the OMB Director waives the requirement. If the requirement is waived, the OMB Director must submit written justification to the House and Senate Appropriations Committees. Grant recipients that receive a waiver are required to return any funds remaining unexpended after 24 months to the awarding federal agency.[77]

Planning for and Projecting Future Vulnerabilities and Risks[78]

The Administration's request proposed that federal agencies

> work in partnership with State, local, and tribal officials to develop mutually agreed upon assessments of future risks and vulnerabilities facing the region, including extreme weather, sea level rise, and coastal flooding and incorporate these into their recovery planning and implementation.[79]

While Congress did not address these specific factors in P.L. 113-2, the language of the request was reflected in the text of Section 1104 of Senate-passed H.R. 1. For example, Section 1104(a) of Senate-passed H.R. 1 would have directed federal agencies, in partnership with state, tribal, and local governments to "*inform* plans for response, recovery, and rebuilding to reduce vulnerabilities from and build long-term resiliency to future extreme weather events, sea level rise, and coastal flooding" (italics added). Further, the provision stated that with respect to "repairing, rebuilding, or restoring infrastructure and restoring land, project sponsors *shall consider, where appropriate*, the increased risks and vulnerabilities associated with future extreme weather events, sea level rise and coastal flooding" (italics added). Section 1104(b) would also have made available funds under the legislation for the coordinated development of "regional projections and assessments of future risks" to help improve the plans required under 1104(a). In general, the impact of this full provision would have depended on how the relevant federal agencies interpreted and implemented the directive to *inform* their plans, and how recipient project sponsors interpreted and implemented the directive to *consider* these increased risks. It is possible, for example, that a requirement, or choice, to take into account the risks delineated in the provision could have resulted in the need for new flood hazard maps that reflect new flood insurance zones based on the future impact of extreme weather events, sea level rise, and coastal flooding; and also possibly new floodplain management standards requiring communities under the NFIP that reflect new land-use planning and construction standards in Special Flood Hazard Areas (SFHA). Also by example, in interpreting and implementing this provision, the U.S. Army Corps of Engineers could have adjusted their plans for the level of flood protection needed along the eastern seaboard.

Mitigation of Future Power Outages[80]

Section 1105 of Senate-passed H.R. 1 as amended would have required the Secretary of HUD, as the chair of the Hurricane Sandy Rebuilding Task Force,[81] to issue guidelines on how recipients of federal funds for reconstruction should "to the greatest extent practicable ... maximize the utilization of technologies designed to mitigate future power outages, continue delivery of vital services and maintain the flow of power to facilities critical to public health, safety and welfare." These guidelines could have been issued in a number of ways, ranging from policy guidance to enforceable regulations. Depending on the scope of the guidelines and whether recipients were required to follow them, the guidelines could have impacted the expenditure of funds for a number of programs. For example, recipients may have been more likely to invest funds received from FEMA's Hazard Mitigation Grant Program or HUD's Community Development Block Grant program in technologies that would mitigate power outages, such as backup generators. The Administration's proposal did not specifically request this provision, but it was arguably consistent with the Administration's emphasis on using funding to mitigate future damages. No similar provision was included in P.L. 113-2.

Embassy Security[82]

Section 1107 of Senate passed H.R. 1 would have authorized the Department of State to transfer up to about $1 billion in Overseas Contingency Operations (OCO) funds, previously appropriated in FY2012 for operations in Iraq, for increased security at U.S. embassies and other overseas posts identified in the Department's security review after the Benghazi attack. These unobligated funds are no longer needed because of reduced operations in Iraq, according to Senator Mikulski. CBO had determined that the amendment had no outlay scoring impact, but the legislation did require the Department of State to follow congressional notification requirements prior to using the funds. P.L. 113-2 carried no such provision, and it was not included in the Administration's formal request.

APPENDIX. SUMMARY OF THE ADMINISTRATION'S REQUEST

The Administration's proposal included $47.44 billion in funding for response and recovery, and $12.97 billion specifically for mitigation of damage from potential future storms and flooding. This division is not typical of recent supplemental requests, and does not conform to either traditional definitions of "recovery and repair" versus "mitigation" or the recent patterns for funding mitigation.

Of note, there are four accounts that have funding requests for both "repair and recovery" and "mitigation," and five accounts where the request for mitigation was the only request.[83] The Administration also requested that the mitigation portion include legislative provisions that would allow monies to be flexibly transferred between programs.

In reading the mitigation portion of the Administration's request, it is useful to understand how the Administration may be defining "recovery and repair" and "mitigation." Using definitions drawn from Presidential Policy Directive 8 (PPD-8), "recovery" refers to

> those capabilities necessary to assist communities affected by an incident to recover effectively, including, but not limited to, rebuilding infrastructure systems; providing adequate interim and long-term housing for survivors; restoring health, social, and community services;

promoting economic development; and restoring natural and cultural resources.

In the same Directive, the Administration noted that "mitigation" refers to

> those capabilities necessary to reduce loss of life and property by lessening the impact of disasters. Mitigation capabilities include, but are not limited to, community-wide risk reduction projects; efforts to improve the resilience of critical infrastructure and key resource lifelines; risk reduction for specific vulnerabilities from natural hazards or acts of terrorism; and initiatives to reduce future risks after a disaster has occurred.[84]

If one relies on these definitions, the key difference between recovery funding and mitigation funding may be that the mitigation funding will be explicitly directed to "initiatives to reduce future risk after a disaster has occurred." However, some of the activities outlined in the Administration's proposal as "mitigation" appeared to be orientated towards "recovery and repair," and vice versa. For example, the Administration proposed $400 million in mitigation funding for the Fish and Wildlife Service's Resource Management account that would be used, among other purposes, for "restoring and enhancing natural systems on State, local and private lands."[85] Further, the Administration's proposal for mitigation funds did not include at least one noteworthy program most traditionally linked with hazard mitigation, that being FEMA's Hazard Mitigation Grant Program (HMGP), which is funded through the Disaster Relief Fund.[86]

When Congress considered supplemental funding for Hurricane Sandy, it did not follow the Administration's format for providing a distinction between funding for "mitigation" and the funding for "repair and recovery." In Senate-passed H.R. 1 there was no distinct chapter or title that separately funded accounts for mitigation. As H.R. 152 worked its way through the House on its way to enactment, $17.1 billion was identified by the House Appropriations Committee as being for "immediate needs." An amendment provided $33 billion in additional funds, but the distinction between the "immediate needs" and other assistance was not clearly split between "repair and recovery" and "mitigation. Provisions in H.R. 152 as it passed the House included "mitigation" as part of the purpose of the funds, without separating that purpose from recovery.[87]

Table A-1. Survey of FY2013 Hurricane Sandy Supplemental Request
millions of dollars in budget authority

Recipient of Appropriation			Request				Recipient Type	
Agency	**Bureau**	**Account**	**Response and Recovery**	**Mitigation**	**Total**	**% of Total Request**	**Federal Damage Recovery**	**Other Assistance**
Agriculture, Rural Development, Food and Drug Administration, and Related Agencies Subcommittee								
Department of Agriculture	Farm Service Agency	Emergency Conservation Program	15.00		15.00	0.02%	No	Yes
	Farm Service Agency	Commodity Credit Corporation Fund	23.00		23.00	0.04%	No	Yes
	Natural Resources Conservation Service	Watershed and Flood Prevention Operations	30.00	150.00	180.00	0.30%	No	Yes
Commerce, Justice, Science and Related Agencies Subcommittee								
Department of Commerce	National Oceanographic and Atmospheric Administration	Operations, Research and Facilities	33.00	360.00	393.00	0.65%	Yes	Yes
	National Oceanographic and Atmospheric Administration	Procurement, Acquisition, and Construction		100.00	100.00	0.17%	No	Yes
Defense Subcommittee								
Department of Defense	Operations and Maintenance	Operations and Maintenance, Navy	41.20		41.20	0.07%	Yes	No
	Revolving and Management Funds	Working Capital Fund, Navy	24.20		24.20	0.04%	Yes	No
Energy and Water Development and Related Agencies Subcommittee								
Army Corps of Engineers - Civil Works	Investigations			30.00	30.00	0.05%	No	Yes
	Construction			3,820.00	3,820.00	6.32%	No	Yes
	Operations and Maintenance		899.00		899.00	1.49%	Yes	Yes
	Flood Control and Coastal Emergencies		592.00		592.00	0.98%	No	Yes

Recipient of Appropriation			Request				Recipient Type	
Agency	Bureau	Account	Response and Recovery	Mitigation	Total	% of Total Request	Federal Damage Recovery	Other Assistance
Financial Services and General Government Subcommittee								
Small Business Administration	Salaries and Expenses		50.00		50.00	0.08%	No	Yes
	Surety Bond Guarantees Revolving Fund				0.00	0.00%	No	Yes
	Disaster Loan Program Account		750.00		750.00	1.24%	No	Yes
Homeland Security Subcommittee								
Department of Homeland Security	Coast Guard	Operating Expenses	66.84		66.84	0.11%	Yes	No
	Coast Guard	Acquisition, Construction and Improvements	207.39		207.39	0.34%	Yes	No
	FEMA	Disaster Relief Fund	11,500.00		11,500.00	19.04%	Yes	Yes
	FEMA	Disaster Assistance Direct Loan Program	300.00		300.00	0.50%	No	Yes
	FEMA	National Flood Insurance Fund	9,700.00		9,700.00	16.06%	No	Yes
Interior, Environment and Related Agencies Subcommittee								
Department of the Interior	U.S. Fish and Wildlife Service	Resource Management		400.00	400.00	0.66%	No	Yes
	U.S. Fish and Wildlife Service	Construction	78.00		78.00	0.13%	Yes	No
	National Park Service	Construction	348.00		348.00	0.58%	Yes	No
Environmental Protection Agency	State and Tribal Assistance Grants			610.00	610.00	1.01%	No	Yes
Labor, Health and Human Services, Education, and Related Agencies Subcommittee								
Department of Labor	Employment and Training Administration	Training and Employment Services	50.00		50.00	0.08%	No	Yes

Table A-1. (Continued)

Recipient of Appropriation			Request				Recipient Type	
Agency	**Bureau**	**Account**	**Response and Recovery**	**Mitigation**	**Total**	**% of Total Request**	**Federal Damage Recovery**	**Other Assistance**
Department of Health and Human Services	Administration for Children and Families	Social Services Block Grant	500.00		500.00	0.83%	No	Yes
	Administration for Children and Families	Children and Families Services Programs	100.00		100.00	0.17%	No	Yes
	Departmental Management	Public Health and Social Services Emergency Fund	200.00		200.00	0.33%	No	Yes
Military Construction, Veterans Affairs and Related Agencies								
Department of Defense—Military Construction	Military Construction, Army National Guard		24.24		24.24	0.04%	Yes	No
Department of Veterans' Affairs	Veterans Health Administration	Medical Services	21.00		21.00	0.03%	Yes	No
	Departmental Administration	Construction, Major Projects	207.00		207.00	0.34%	Yes	No
Transportation, Housing and Urban Development, and Related Agencies								
Department of Transportation	Federal Aviation Administration	Facilities and Equipment	30.00		30.00	0.05%	Yes	No
	Federal Highway Administration	Emergency Relief Program	308.00		308.00	0.51%	No	Yes
	Federal Railroad Administration	Operating Subsidy Grants to NRPC	32.00		32.00	0.05%	No	Yes
	Federal Transit Administration	Public Transportation Emergency Relief Program	6,200.00	5,500.00	11,700.00	19.37%	No	Yes

Recipient of Appropriation			Request				Recipient Type	
Agency	**Bureau**	**Account**	**Response and Recovery**	**Mitigation**	**Total**	**% of Total Request**	**Federal Damage Recovery**	**Other Assistance**
Department of Housing and Urban Development	Public and Indian Housing Programs	Tenant-Based Rental Assistance			N/A	N/A	No	Yes
	Community Planning and Development	Community Development Fund	15,000.00	2,000.00	17,000.00	28.14%	No	Yes
Requests for Appropriations of $10 Million or Less								
30 Individual requests of $10 million or less			108.80		108.80	0.18%	88.07	19.73
		TOTAL	**47,438.67**	**12,970.00**	**60,408.67**	**100.00%**		

Source: CRS analysis of the FY2013 Supplemental Appropriations Request, as transmitted in a letter from Jeffrey D. Zients, Deputy Director for Management, to The Honorable John Boehner, Speaker of the House of Representatives, December 7, 2012.

In another circumstance, provisions in Senate-passed H.R. 1 specifically identified subset of funds from the total appropriation to an account that may be used exclusively for "mitigation."[88] In addition, Sections 1104 and 1105 of Senate-passed H.R. 1, which were general provisions applying to all funds in the legislation, would have encouraged funds provided in the bill to be used in a manner that mitigates future risks.

Requested funding levels are provided by appropriations account in **Table A-1**, below. It provides a summary and brief analysis of the Administration's budget request. A series of columns notes the agency, bureau, and account for which appropriations were requested. The table then notes how much the Administration sought as funds needed for recovery and repair of damage, as opposed to mitigation of future disaster impacts, and a total of the two categories. The table then notes what percentage that request is of the overall total sought. Finally the table includes a quick assessment of whether the appropriation is intended to pay for damaged federal government property or provide other disaster assistance. These final categories on potential recipients are not mutually exclusive at the account level. Requests for appropriations of $10 million or less are combined in a single line for the sake of brevity—as the table indicates, these 30 items represent less than 0.2% of the total request. They can be found in the more complete accounting of the request, and the Senate legislative response to date, in **Table 1**.

End Notes

[1] Federal Emergency Management Agency, "Hurricane Sandy: Timeline," http://www.fema.gov/hurricane-sandy-timeline.

[2] Voiland, Adam, "Comparing the Winds of Sandy and Katrina," November 9, 2012, http://www.nasa.gov/mission_pages/hurricanes/archives/2012/h2012_Sandy.html.

[3] As downloaded from http://www.eqecat.com/news/in-the-news/2012/hurricane-sandy/, link verified January 23, 2013.

[4] The Stafford Act is codified at 42 U.S.C. 5121 et seq. To date, the major disaster declarations are: New York (DR-4085); New Jersey (DR-4086); Connecticut (DR-4087); Rhode Island (DR-4089); Delaware (DR-4090); Maryland (DR-4091); Virginia (DR-4092); West Virginia (DR-4093); New Hampshire (DR-4095); the District of Columbia (DR-4096); Massachusetts (DR-4097); Ohio (DR-4098); and Pennsylvania (DR-4099). More information on eachdeclaration is available at http://www.fema.gov/disasters.

[5] Office of Management and Budget, *Hurricane Sandy Funding Needs*, Washington, DC, December 7, 2012, http://www.whitehouse.gov/sites/default/files/ supplemental__december_7_2012_hurricane_sandy_funding_needs.pdf.pdf.

[6] H.R. 1 was a continuing resolution for FY2011 passed by the House in the 112th Congress that was not previously voted on in the Senate.

[7] Slight changes were made from the draft on the Senate website including designating what had been chapters as titles, and altering the section numbering.

[8] For more information, see CRS Report R42037, *SBA Surety Bond Guarantee Program*, by Robert Jay Dilger.

[9] For more information, see CRS Report R42850, *The National Flood Insurance Program: Status and Remaining Issues for Congress*, by Rawle O. King.

[10] The analysis in this report of the House position is based on those texts from the House Rules Committee website.

[11] P.L. 112-25.

[12] The BCA also specifically redefined "disaster relief" as being federal government assistance provided pursuant to a major disaster declared under the Stafford Act, not to be confused with funding provided for other types of incidents, or exclusively resources provided through the Disaster Relief Fund (DRF).

[13] For a more extensive discussion of this structure, see CRS Report R42352, *An Examination of Federal Disaster Relief Under the Budget Control Act*, by Bruce R. Lindsay, William L. Painter, and Francis X. McCarthy.

[14] The emergency designation for the Army Corps of Engineers Construction account was stricken by a point of order on the Senate floor. See *Congressional Record,* December 21, 2012, pp. S8341-S8342.

[15] CBO, "Estimate of the Disaster Relief Appropriations Act, 2013 (H.R. 152) as Cleared by Congress for the President's Signature on January 28, 2013," January 29, 2013. The total score against the discretionary budget cap is $2million lower due to conversion of some unobligated balances of budget authority to emergency funding in the bill.

[16] This section prepared by Megan Stubbs, Specialist in Agricultural Conservation and Natural Resources Policy (7-8707) and Randy Aussenberg, Analyst in Nutrition Assistance Policy (7-8641).

[17] USDA, "Farmers and Ranchers Urged to Record Losses from Hurricane Sandy," press release, October 31, 2012, http://www.usda.gov/wps/portal/usda/ usdahome?contentid=2012/10/0337.xml&contentidonly=true.

[18] USDA, "USDA Delivers Funding for Hurricane Sandy Recovery Projects in 11 States," press release, November 8, 2012, http://www.usda.gov/wps/portal/usda/ usdahome?contentid=2012/11/0342.xml&contentidonly=true.

[19] Additional information on ECP, EFRP, EWP, and EWP floodplain easements—collectively referred to as emergency agricultural land assistance programs—may be found in CRS report, CRS Report R42854, *Emergency Assistance for Agricultural Land Rehabilitation.*

[20] Aside from TEFAP, many of the food assistance benefits provided and being provided by USDA's Food and Nutrition Service programs (such as the Disaster Supplemental Nutrition Assistance Program (D-SNAP)) require no additional appropriations because the benefits are entitlements.

[21] Prepared by Nathan James, Analyst in Crime Policy (7-0264).

[22] In addition to the fisheries failure that was declared for New Jersey and New York fisheries, during 2012 disasters were also declared for Alaska Chinook salmon, New England groundfish, Mississippi fisheries, and American Samoa bottomfish.

[23] The amendment to H.R. 152 offered by Congressman Frelinghuysen (H.Amdt. 5) would have provided a total of $290.0 million for the ORF account, which included $150.0 million for Regional Ocean Partnership grants to coastal states impacted by Hurricane Sandy. However, the House adopted an amendment offered by Congressman Flores (H.Amdt. 6), which struck the $150.0 in funding for Regional Ocean Partnership grants and reduced funding for the ORF account to $140.0 million.

[24] Prepared by Charles Stern, Specialist in Natural Resources Policy (7-7786), and Nicole Carter, Specialist in Natural Resources Policy (7-0854).

[25] The Administration's request for the Corps included account-level funding requests and descriptions; it did not include bill language, which complicates comparisons with some of the House and Senate provisions.

[26] Supplemental appropriations for the Corps were proposed for five accounts: the Investigations account for new and ongoing Corps studies; the Flood Control and Coastal Emergencies (FCCE) account for flood fighting, preparedness and response, and repair of eligible damaged nonfederal flood and hurricane protection projects; the Operations and Maintenance (O&M) account for operational Corps projects; the Construction account for construction of new projects or major upgradess; and the General Expenses account for administrative and oversight.

[27] The Administration request did not specify whether the Corps funding was an "emergency requirement." For more information on this designation, see above section, "Disaster Relief and Emergency Funding Under the Budget Control Act."

[28] The Administration used the term "mitigation" for Corps construction projects. Most Corps projects reduce flood risk by reducing the vulnerability to the flood hazard (i.e., structures that reduce the probability of an area flooding), not by reducing the consequence if a flood event occurs (i.e., limits the value of the damaged property). Typically it is the latter type of activity that has been referenced to as "mitigation" among federal programs and floodplain mangers.

[29] The Senate bill did not define "previous natural disasters" or further spell out the terms for use of this funding.

[30] Nonfederal cost shares, as specified in statute, are 35% for Corps flood and coastal storm damage reduction projects and 50% for beach renourishment components projects that have been authorized since 2003. Notably, in those cases nonfederal LEERD costs are counted toward the nonfederal share.

[31] The Administration Request also proposed a 90/10 cost share, but did not provide directions on the treatment of LEERD costs.

[32] While not specified in the bill, all other construction projects that are not "ongoing" potentially would be subject to the typical cost sharing requirements referenced above. This would exclude from the cost-share waiver later beach renourishment activities for coastal storm damage reduction projects carried out with funds other than those provided in the supplemental legislation.

[33] Many repairs to existing coastal storm damage reduction projects are eligible for 100% funding under the Corps FCCE account for repair to their pre-storm conditions. Improvements that go beyond repair would not be eligible for FCCE funding, and would have to be funded by the Construction account.

[34] Specifically, receipt of these funds is contingent upon the Corps completing and providing to Congress an interim report that includes an assessment of authorized Corps projects for reducing flooding and storm risks in the area affected by the storm that have been constructed or are under construction. The deadline for this report is March 1, 2013.

[35] GSA component prepared by Garrett Hatch, Specialist in American National Government, 7-7822. SBA component prepared by Bruce Lindsay, Analyst in American National Government, 7-3752, and Robert Dilger, Senior Specialist in American National Government, 7-3110.

[36] Senate-passed H.R. 1 would have provided $40 million for salaries and expenses of which, $20 million was for grants or cooperative agreements for public-private partnerships to provide economic development assistance to industries and/or regions affected by Hurricane Sandy.

[37] The DRF provides funding for the majority of disaster assistance programs authorized under the Stafford Act (42 U.S.C. 5121 et seq), including Public Assistance, Individual Assistance, and Hazard Mitigation Assistance. For more, see CRS Report RL33053, *Federal Stafford Act Disaster Assistance: Presidential Declarations, Eligible Activities, and Funding*, by Francis X. McCarthy.

[38] For more on the Community Disaster Loan program (Sec. 417 of the Stafford Act), see CRS Report R42527, *FEMA's Community Disaster Loan Program: History, Analysis, and Issues for Congress*, by Jared T. Brown.

[39] Prepared by Rawle O. King, Specialist in Financial Economics and Risk Assessment, 7-5975.

[40] P.L. 109-65; 110 Stat. 1998.

[41] P.L. 109-106; 119 Stat. 2288.

[42] P.L. 109-208; 120 Stat. 317.

[43] Under current law, FEMA must repay any borrowed funds (with interest) as it collects premiums. However, FEMA is unlikely to repay the funds borrowed to pay 2005 hurricane-related claims within the next 10 years. Even if FEMA increased flood insurance rates up to the maximum amount allowed by law (20% per year), the program would still not have sufficient funds to cover future obligations for policyholder claims, operating expenses, and interest on debt stemming from recent catastrophic flood events. Some experts have suggested that Congress consider forgiving some or all of NFIP's Treasury borrowing. Supporters of debt forgiveness point to billions of dollars in flood losses that would otherwise have been paid by the Treasury and thus taxpayers. According to FEMA, the NFIP saves taxpayers over $1.7 billion annually in flood losses that, in the absence of the program, would be paid by taxpayers. Debt forgiveness could, however, be judged an explicit subsidy from general taxpayer funds, with federal budgetary consequences.

[44] The Biggert-Waters Flood Insurance Reform Act of 2012 (P.L. 112-141) includes provisions to: (1) phase out long-running premium subsidies for vacation homes, businesses, and repetitive loss properties (those that have made repeated claims on the program); (2) direct FEMA to include catastrophic loss years when assessing flood risk in order to set annual premium rates; and (3) establish a reserve fund to offset claims during catastrophic loss years.

[45] For more on the NFIP, see CRS Report R42850, *The National Flood Insurance Program: Status and Remaining Issues for Congress*, by Rawle O. King.

[46] For information and support on the Disaster Recovery Act of 2012 or the Sandy Recovery Improvement Act of 2013, please contact any of the following analysts: Francis X. McCarthy, fmcarthy@crs.loc.gov, 7-9533; Jared T. Brown,jbrown@crs.loc.gov, 7-4918; and Edward C. Liu, eliu@crs.loc.gov, 7-9166.

[47] Section 404 of the Stafford Act.

[48] Sections 403(a)(3)(A), 406, 407, and 502(a)(5) of the Stafford Act. Respectively these sections of the Stafford Act refer to essential assistance debris removal; repair, restoration, and replacement of damaged facilities; non-essential debris removal; and debris removal (via an emergency declaration).

[49] Section 408 of the Stafford Act.

[50] Section 417 of the Stafford Act.

[51] For details on this program, see CRS Report RL34146, *FEMA's Disaster Declaration Process: A Primer*, by Francis X. McCarthy.

[52] Defined by the Census Bureau as Iowa, Kansas, Minnesota, Missouri, Nebraska, North Dakota, and South Dakota.

[53] Prepared by Carol Hardy-Vincent, Specialist in Natural Resources Policy, 7-8651.

[54] Prepared by David Bradley, Specialist in Labor Economics, 7-7352.

[55] For more information, see CRS Report R41135, *The Workforce Investment Act and the One-Stop Delivery System*, by David H. Bradley.

[56] Prepared by Karen Lynch, Specialist in Social Policy, 7-6899 and Sarah A. Lister, Specialist in Public Health and Epidemiology, 7-7320.

[57] The request did not specify when requested funds should be available. Senate-passed H.R. 1 would have made SSBG funds available for two succeeding fiscal years; Head Start program funds available through FY2014; and PHSSEF funds available until expended.

[58] For more information, see CRS Report 94-953, *Social Services Block Grant: Background and Funding*, by Karen E. Lynch.

[59] For more information, see CRS Report RL30952, *Head Start: Background and Issues*, by Karen E. Lynch.

[60] Senate Appropriations Committee, "Summary: Fiscal Year 2013 Disaster Assistance Supplemental," press release, December 12, 2012, http://www.appropriations.senate.gov, p. 7.

[61] H.R. 1, EAS, p. 83, and H.R. 152, EH, p. 21.

[62] Note that while Title 8 of H.R. 152 also included appropriations for the Departments of Transportation and Housing and Urban Development, Section 1094 provides that Title 10 Chapter 9 shall apply in place of Title 8.

[63] Prepared by D. Randy Peterman, Analyst in Transportation Policy, 7-3267.

[64] For additional background on this issue, see CRS Report R42804, *Emergency Relief Program: Federal-Aid Highway Assistance for Disaster-Damaged Roads and Bridges*, by Robert S. Kirk.

[65] Prepared by Maggie McCarty, Specialist in Housing Policy, 7-2163, and Eugene Boyd, Analyst in Federalism and Economic Development Policy, 7-8689.

[66] Prepared by Natalie M. Keegan, Analyst in American Federalism and Emergency Management Policy, nkeegan@crs.loc.gov, 7-9569.

[67] Internal controls are measures that the federal agency takes to ensure that the federal agency and grant recipients are in compliance with applicable statutes, regulations, and OMB circulars. Internal control standards seek to ensure that the use of funds comply with applicable laws, that assets are appropriately protected against waste, fraud, and abuse, and that federal agencies have efficient and effective financial and program administration systems that allow for appropriate accountability of funds.

[68] For example, there are statutory provisions for single audit reporting are contained in the Single Audit Act of 1984 (P.L. 98-502), as amended, and OMB Circular A-133, *Audits of States, Local Governments, and Non-Profit Organizations*. Examples of statutory provisions for improper payments can be found in the Improper Payments Act of 2002 (P.L. 107-300). Examples of internal control provisions can be found in the Federal Managers' Financial Integrity Act of 1982, as codified in 31 U.S.C. 3512, with OMB guidance contained in Circular A-123, *Managements Responsibility for Internal Controls*.

[69] 112th Congress, H.R. 1, section 1103(a).

[70] Examples of funding provided for oversight of grant funds include $1 million for the Environmental Protection Agency's State and Tribal Assistance Grants for management and oversight, and an allowance for a percentage (less than 1%) of funds provided to the Federal Transit Administration's Public Transportation Emergency Relief Program and the Federal Railroad Administration's National Railroad Passenger Corporation grant to be used for management and oversight.

[71] H.R. 152, EH, section 904(a)(1) and 904(a)(2).

[72] H.R. 152, EH, section 904(b).

[73] P.L. 107-300, *Improper Payments Information Act of 2002*, as amended by P.L. 111-204, *Improper Payments Elimination and Recovery Act of 2010*.

[74] 31 U.S.C. 3321.

[75] H.R. 1, EAS, 112th Congress, Sec. 1109. Funeral costs were exempted from this provision, though funeral costs are traditionally funded under the "other needs assistance" provisions of the Stafford Act and are provided to the surviving family member rather than to the deceased individual (42 U.S.C. 5174).

[76] 112th Congress, H.R. 1, as amended, section 1108.

[77] H.R. 152, as amended, Section 904(c).

[78] Prepared by Jared Brown, Analyst in Emergency Management Policy, 7-4918.

[79] Office of Management and Budget, *Hurricane Sandy Funding Needs*, Washington, DC, December 7, 2012, second page of *Appendix: Detailed Estimates of Necessary Federal Resources*.

[80] Prepared by Jared Brown, Analyst in Emergency Management Policy, 7-4918.

[81] The Secretary of HUD was designated as the chair in Executive Order 13632, "Establishing the Hurricane Sandy Rebuilding Task Force," 77 *Federal Register* 74341, December 14, 2012,

https://www.federalregister.gov/articles/2012/12/14/2012-30310/establishing-the-hurricane-sandy-rebuilding-task-force.

[82] Prepared by Susan B. Epstein, Specialist in Foreign Policy, 7-6678.

[83] The accounts receiving requests for both "recovery and repair" and "mitigation" are: Dept. of Agriculture's Watershed and Flood Prevention Operations; NOAA's Operations, Research and Facilities; Federal Transit Administration's Public Transportation Emergency Relief Program; and HUD's Community Development Fund.

[84] Office of Management and Budget, *Hurricane Sandy Funding Needs*, Washington, DC, December 7, 2012, p. 69.

[85] White House, *Presidential Policy Directive 8: National Preparedness*, Washington, DC, March 30, 2011, p.6, http://www.dhs.gov/presidential-policy-directive-8-national-preparedness.

[86] The Administration does request $11.5 billion for the Disaster Relief Fund (DRF) in the "recovery and repair" section of the request, of which some to be determined amount will be used for HMGP. The amount of assistance provided through HMGP is set through a statutory formula per disaster declaration. For more on this program, see CRS Report R40471, *FEMA's Hazard Mitigation Grant Program: Overview and Issues*, by Natalie Keegan.

[87] For example, see the designation of $336 million in budget authority for the National Railroad Passenger Corporation in Chapter 10.

[88] For example, see the designation of $2 billion in budget authority for the Community Development Fund in Chapter 10.

INDEX

A

abuse, 24, 76, 92
access, 79
accountability, 17, 92
accounting, 88
adjustment, 6, 50, 51, 64
Administration for Children and Families, 47, 70, 71, 86
administrators, 77
ADR, 22, 23, 24, 35, 36
advancement, viii, 2
agencies, 7, 17, 19, 23, 34, 54, 57, 65, 67, 69, 71, 72, 74, 77, 78, 79, 80, 92
Air Force, 44
Alaska, 6, 31, 89
Alternative Dispute Resolution, 35
American Recovery and Reinvestment Act, 22, 35
American Recovery and Reinvestment Act of 2009, 35
American Samoa, 89
APA, 23
appropriations, vii, ix, 1, 3, 4, 24, 40, 41, 42, 43, 50, 51, 53, 56, 57, 58, 62, 65, 67, 68, 69, 71, 72, 76, 77, 78, 79, 88, 89, 90, 92
Appropriations Act, vii, ix, 1, 3, 4, 28, 32, 39, 89
Appropriations Committee, 41, 42, 59, 67, 78, 80, 83, 92
arbitration, viii, 2, 22, 23, 24, 27, 35
assessment, 25, 29, 55, 88, 90
assets, 31, 34, 35, 54, 55, 60, 67, 92
audit, 16, 78, 92
audits, 64
authorities, vii, 1, 4, 14, 15, 16, 33, 77
authority, viii, ix, 1, 3, 6, 8, 9, 11, 12, 13, 16, 20, 21, 22, 24, 25, 32, 33, 35, 39, 41, 42, 43, 49, 50, 51, 59, 61, 62, 69, 75, 76, 77, 84, 89, 93

B

banks, 53
base, viii, 2, 15, 18, 34, 71
benefits, 15, 18, 76, 89
breakdown, 42
budget resolution, 50
businesses, 76, 91

C

Census, 91
CFR, 31
challenges, 6, 8
childhood, 71
children, 8, 71
City, 74
civil rights, 36, 76
Clean Water Act, 19, 66, 67

clothing, 9
clusters, 19
Coast Guard, 45, 49, 61, 85
coastal communities, 40, 54
collateral, 60
Committees on Appropriations, 77
commodity, 53
communities, 5, 7, 10, 15, 17, 36, 64, 81, 82
community, 8, 17, 19, 33, 76, 82, 83
community service, 17, 82
compliance, 19, 67, 79, 92
composition, 13, 23
Congress, vii, viii, ix, x, 1, 2, 3, 4, 7, 10, 16, 17, 25, 28, 29, 31, 32, 33, 34, 36, 37, 39, 40, 41, 43, 44, 45, 46, 47, 48, 50, 56, 61, 62, 64, 65, 71, 74, 77, 78, 79, 80, 83, 88, 89, 90, 91, 92
construction, 26, 51, 57, 58, 64, 68, 71, 81, 90
Consumer Price Index, 34, 36
cooperation, 7
cooperative agreements, 59, 60, 90
coordination, 32, 55
cost, ix, 2, 4, 6, 9, 10, 11, 12, 13, 14, 15, 17, 18, 20, 23, 31, 32, 33, 36, 41, 52, 57, 58, 59, 60, 64, 76, 90
CPI, 17
critical infrastructure, 83
cultural values, 67
culture, 67
Customs and Border Protection, 45, 61

D

damages, vii, ix, 3, 15, 33, 39, 41, 57, 74, 81
deaths, ix, 8, 39, 40
decision-making process, 8
denial, 35
Department of Agriculture, 84
Department of Commerce, 84
Department of Defense, 56, 84, 86
Department of Energy, 40
Department of Health and Human Services, 69, 86
Department of Homeland Security, 16, 31, 32, 33, 34, 35, 36, 61, 85
Department of Justice, 54
Department of Labor, 36, 68, 85
Department of the Interior, 65, 85
Department of Transportation, 72, 73, 86
destruction, 60
DHS, 21, 31, 61, 64
directives, 18
disaster, vii, viii, ix, 1, 2, 3, 4, 5, 6, 7, 8, 9, 10, 13, 14, 16, 19, 20, 21, 25, 26, 27, 28, 29, 31, 32, 33, 34, 35, 39, 40, 41, 42, 50, 51, 52, 53, 59, 60, 61, 64, 69, 71, 73, 74, 75, 76, 77, 79, 80, 83, 88, 89, 90, 93
disaster area, 9, 19, 61
disaster assistance, vii, viii, ix, 1, 2, 3, 4, 5, 19, 21, 28, 39, 40, 41, 42, 79, 88, 90
disaster assistance programs, vii, ix, 1, 3, 4, 28, 40, 42, 90
disaster relief, 4, 6, 13, 50, 51, 61, 89
Disaster Relief Appropriations Act, vii, ix, 1, 3, 4, 28, 39, 89
Disaster Relief Fund, 4, 24, 29, 30, 45, 61, 83, 85, 89, 93
distribution, 56, 69, 75
District of Columbia, ix, 39, 40, 41, 88
DOI, 65, 67
DOJ, 54
DOL, 68
DOT, 72, 77
draft, 20, 36, 41, 71, 77, 88
DRF, 4, 30, 51, 61, 62, 89, 90, 93
drinking water, 66
drought, 52

E

economic development, 83, 90
economic losses, 40
ecosystem, 66
elementary school, 15, 33
emergency, viii, 2, 4, 5, 6, 7, 8, 17, 18, 27, 31, 33, 34, 40, 50, 51, 52, 53, 56, 58, 59, 61, 69, 72, 74, 78, 89, 90, 91

Emergency Assistance, vii, 1, 3, 5, 35, 41, 89
emergency declarations, viii, 2, 40
emergency management, 5, 6, 7, 69
emergency relief, 72, 74
Emergency Relief Program, 48, 72, 73, 74, 86, 92, 93
emergency response, 59
employees, viii, 2, 15, 18, 34, 70
employment, 68
environmental impact, 20
Environmental Protection Agency, 65, 85, 92
environmental resources, 19
EPA, 65, 66
equipment, 54, 55, 56, 61, 72, 73, 74
erosion, 52
exclusion, 33
Executive Order, 37, 92
exercise, 33
expenditures, 10, 19, 58
extreme weather events, 80

F

FAA, 73
families, 71
family units, 9
FBF, 59
federal agency, 23, 78, 79, 80, 92
federal assistance, 11, 14, 23, 33
Federal Buildings Fund, 45, 59
Federal Bureau of Investigation, 43
Federal Emergency Management Agency, vii, 1, 31, 32, 33, 34, 35, 36, 40, 45, 88
federal funds, 81
federal government, 5, 9, 10, 11, 15, 17, 50, 88, 89
Federal Highway Administration, 48, 74, 86
federal law, 19
Federal Register, 31, 33, 34, 37, 92
federal regulations, 70
FEMA, vii, viii, ix, 1, 2, 4, 5, 7, 8, 9, 10, 11, 12, 13, 14, 15, 16, 17, 18, 20, 21, 22, 23, 24, 25, 26, 28, 29, 31, 32, 33, 34, 35, 36, 37, 40, 62, 64, 70, 71, 76, 81, 83, 85, 90, 91, 93
financial, 10, 15, 16, 29, 60, 62, 75, 78, 92
financial incentives, 16
financial resources, 62
Fish and Wildlife Service, 46, 65, 83, 85
fisheries, 89
flexibility, 6, 11, 20, 26, 33, 53, 74
flooding, 10, 54, 57, 72, 80, 82, 90
floods, 52, 63
food, 53, 89
Food and Drug Administration, 43, 51, 84
force, 37, 40, 93
forecasting, 40, 55
formula, 66, 70, 71, 93
fraud, 76, 92
funding, ix, x, 4, 20, 39, 40, 41, 42, 48, 50, 51, 52, 53, 54, 56, 57, 58, 59, 60, 61, 65, 67, 68, 69, 70, 72, 73, 74, 76, 77, 78, 79, 81, 82, 83, 88, 89, 90, 92
funds, viii, 2, 13, 14, 19, 29, 32, 33, 48, 49, 50, 52, 53, 56, 57, 58, 59, 61, 63, 64, 66, 67, 68, 69, 70, 71, 74, 75, 76, 77, 78, 79, 80, 81, 82, 83, 88, 90, 91, 92

G

GAO, 37, 78
General Services Administration, 59
General Services Administration (GSA), 59
governments, 5, 6, 7, 10, 16, 18, 29
governor, viii, 2, 5, 6, 64
grant programs, 78, 79
grants, 10, 11, 12, 15, 17, 32, 33, 34, 59, 60, 66, 67, 77, 78, 79, 89, 90
GSA, 59, 90
guidance, 13, 19, 31, 33, 34, 35, 69, 78, 79, 81, 92
guidelines, 6, 13, 28, 81
Gulf Coast, vii, 3, 35, 57

H

habitat, 67

Hazard Mitigation Grant Program, viii, 2, 3, 7, 19, 35, 63, 64, 81, 83, 93
hazards, 17, 52
health, 68, 69, 70, 82
Health and Human Services, 47, 68, 85
health care, 70
health services, 70
HHS, 49, 68, 69, 70, 71
high school, 33
history, ix, 3, 20, 26, 36, 67
homeland security, 62, 63
homes, 10, 91
host, 34
House, ix, x, 4, 24, 28, 29, 30, 31, 32, 36, 37, 39, 40, 41, 42, 48, 49, 50, 57, 59, 61, 62, 74, 77, 78, 80, 83, 87, 88, 89
House Committee on Transportation and Infrastructure, 24, 31
House of Representatives, ix, 37, 39, 40, 48, 87
housing, viii, 2, 9, 19, 76, 82
Housing and Urban Development, 36, 48, 72, 74, 86, 87, 92
HUD, 72, 74, 75, 76, 77, 81, 92, 93
human, vii, 1
Hurricane Katrina, 3, 8, 10, 31, 37, 62, 64
Hurricane Sandy, vii, ix, 1, 3, 26, 29, 37, 39, 40, 53, 58, 59, 60, 63, 66, 70, 73, 75, 81, 84, 88, 89, 92, 93
hurricanes, 88

I

Immigration and Customs Enforcement, 45
improvements, 10, 12, 42, 55, 74
income, 63, 71, 77
independence, 5
Indian Affairs Committee, 5
individuals, 10, 18, 79
industries, 90
inflation, 22
information technology, 60
infrastructure, 42, 55, 67, 73, 74, 80, 82
injuries, 8, 25
injury, 35, 60
integrity, 62, 78
interim regulations, 73
Iowa, 10, 91
Iraq, 82
islands, 68
isolation, 6
issues, 12, 20, 29, 77

J

job training, 68
jurisdiction, 16, 23, 27, 50, 51
justification, 80

L

language barrier, 6
laws, 5, 27, 69, 92
laws and regulations, 27
legislation, vii, ix, 1, 3, 16, 27, 28, 39, 40, 41, 42, 50, 61, 62, 64, 72, 78, 79, 80, 82, 88, 90
levees, 64
light, 22
litigation, 35, 36
loans, 29, 60, 62, 66
local government, 5, 6, 10, 14, 15, 17, 18, 25, 29, 31, 32, 33, 40, 57, 61, 62, 64, 67, 74, 75, 76, 80
Louisiana, 31, 57, 64

M

magnitude, 7
major disaster declaration, ix, 6, 39, 52, 62, 66, 67, 88
majority, 22, 29, 68, 72, 90
management, 15, 20, 21, 60, 78, 81, 92
manufactured housing, 9
mapping, 35, 54, 55
Maryland, 88
materials, 24
matter, 4
medical, 8

mental health, 8, 68, 69, 70, 71
military, 71
mission, 88
missions, 54
Missouri, 91
modifications, 4, 25

N

National Aeronautics and Space Administration, 54
National Defense Authorization Act, 41, 60
National Environmental Policy Act (NEPA), 35
National Flood Insurance Program, viii, ix, 1, 3, 40, 42, 49, 62, 89, 91
National Institutes of Health, 69
National Park Service, 46, 65, 85
national strategy, ix, 2, 4, 25
Native Americans, 6
natural disaster, vii, 3, 8, 52, 57, 58, 66, 67, 68, 90
natural disasters, 57, 58, 67, 68, 90
natural hazards, 83
natural resources, 54
Natural Resources Conservation Service, 43, 52, 84
Natural Resources Conservation Service (NRCS), 52
New England, 89
NOAA, 54, 55, 93
NPS, 65, 67
nutrition, 51

O

Office of Management and Budget, 78, 88, 92, 93
Office of the Inspector General, 43, 45, 47, 48, 49, 61, 69, 75
officials, 21, 24, 35, 57
OMB, 78, 79, 80, 92
operations, 58, 82
oversight, 59, 73, 78, 90, 92
overtime, 15, 18

P

penalties, 76
PHAs, 77
police, 15, 33
policy, 5, 7, 12, 13, 14, 15, 16, 18, 28, 31, 33, 34, 35, 81, 93
politics, 35
population, vii, 3, 33
population density, vii, 3
precedent, 13, 19
preparedness, 40, 54, 90, 93
preservation, 20
President, viii, 2, 6, 9, 12, 17, 18, 19, 20, 21, 27, 32, 34, 35, 40, 41, 42, 43, 44, 45, 46, 47, 48, 51, 52, 53, 56, 58, 59, 61, 62, 64, 65, 66, 68, 69, 70, 71, 72, 73, 74, 77, 78, 79, 89
prevention, 36, 52
private sector, 25, 40
probability, 90
procurement, 75, 76
producers, 52
profit, 8, 14, 32, 33, 35, 76
program administration, 92
project, 10, 11, 12, 13, 14, 15, 17, 20, 22, 32, 33, 35, 36, 56, 57, 58, 59, 72, 76, 77, 80
project sponsors, 80
protection, 36, 52, 81, 90
public assistance, 11, 12, 21, 24, 26, 35
public health, 34, 69, 81
public housing, 77
public officials, 19
public welfare, 14, 33
public-private partnerships, 90

Q

qualifications, 13

R

radio, 54
ramp, 40
reading, 82
real property, 59, 70
recommendations, 16, 25, 29, 33
reconstruction, 14, 15, 68, 81
recovery, viii, 2, 3, 6, 9, 10, 17, 19, 25, 29, 35, 36, 42, 50, 52, 59, 60, 69, 70, 71, 74, 75, 76, 80, 82, 83, 88, 93
recovery plan, 80
recovery process, 9, 25
recycling, 15, 16
Reform, 4, 91
reforms, 3, 4, 28, 36
regulations, 7, 12, 13, 17, 20, 21, 22, 24, 26, 28, 29, 31, 32, 33, 34, 35, 36, 37, 69, 75, 81, 92
rehabilitation, 57
reimburse, 17, 18, 33
relief, 50, 51
remediation, 55
repair, viii, 2, 9, 10, 11, 14, 15, 31, 33, 35, 54, 55, 56, 57, 58, 59, 68, 70, 72, 73, 74, 82, 83, 88, 90, 91, 93
requirements, 11, 14, 19, 20, 21, 23, 27, 36, 60, 76, 82, 90
research facilities, 70
resilience, 83
resolution, viii, 2, 22, 23, 27, 35, 50, 64, 88
resources, viii, 2, 8, 20, 29, 40, 51, 78, 83, 89
response, vii, ix, 1, 3, 6, 10, 21, 36, 40, 42, 50, 52, 59, 60, 62, 69, 71, 80, 82, 88, 90
restoration, viii, 2, 11, 14, 31, 55, 66, 91
restrictions, 9
revenue, 29
rights, 27, 36, 58
risk, 19, 57, 64, 65, 66, 73, 83, 90, 91
risks, 17, 54, 57, 58, 80, 83, 88, 90
rules, 5, 71, 75
runoff, 52

S

SAFE Port Act, 33
safety, 34, 81
salmon, 89
Sandy Recovery Improvement Act of 2013, vii, viii, 1, 2, 3, 4, 5, 7, 8, 9, 11, 13, 14, 16, 17, 18, 19, 22, 25, 28, 29, 30, 31, 36, 63, 91
saving lives, 34
savings, 10, 16, 20
school, 15, 17, 33
scope, 21, 26, 63, 81
sea level, 80
Secret Service, 45
Secretary of Agriculture, 67
security, 60, 82
Security and Accountability For Every Port Act, 33
services, 8, 17, 29, 32, 36, 55, 68, 70, 71, 76, 81
shortage, 29
small businesses, 59, 60
SNAP, 89
social services, 68, 70
South Dakota, 91
sovereignty, 5, 6
specialists, 22
spending, 50, 67, 78
stability, 57
staffing, 60
Stafford Act, vii, viii, 1, 2, 3, 5, 6, 7, 8, 9, 11, 12, 13, 14, 15, 16, 17, 19, 20, 21, 22, 23, 28, 29, 30, 31, 32, 33, 34, 35, 41, 52, 53, 63, 64, 66, 67, 73, 75, 88, 89, 90, 91, 92
state, viii, 2, 5, 6, 7, 8, 9, 10, 14, 15, 17, 18, 19, 20, 21, 25, 29, 31, 32, 33, 34, 40, 52, 53, 55, 64, 66, 67, 74, 75, 76, 77, 80
states, ix, 4, 5, 6, 7, 8, 20, 21, 26, 31, 32, 39, 40, 41, 53, 55, 57, 66, 67, 68, 70, 71, 75, 76, 89
statutes, 92
statutory authority, 63
statutory provisions, 92

stock, 9
storage, 31
storms, vii, 3, 54, 57, 67, 82
structure, 56, 61, 89
subsidy, 61, 62, 91
substance abuse, 69
Superfund, 46
Supplemental Nutrition Assistance Program, 89
surplus, 63
survivors, 82
sustainability, 57

T

target, 66
Task Force, 29, 37, 81, 92
taxpayers, 91
technical assistance, 54, 55, 59, 60, 71, 75
technologies, 81
temporary housing, 8, 9, 27
territory, 5
terrorism, 83
testing, 78
The National Commission on Children and Disasters, 31
threats, 34, 35
time frame, 26, 80
Title V, 4
training, 68
transactions, 27, 36
transportation, 9, 73
trauma, 7, 8
Treasury, 63, 91
treatment, 18, 50, 66, 67, 90
tribal lands, 6
tribal officials, 80
Trust Fund, 46, 73

U

U.S. Army Corps of Engineers, 56, 81
United, ix, 25, 39, 40
United States, ix, 25, 39, 40
USDA, 52, 53, 89

V

victims, 9, 10
vote, ix, 28, 40, 41, 42
vulnerability, 66, 90

W

wages, viii, 2, 15, 16, 18, 33, 70
waiver, 6, 21, 23, 58, 75, 79, 80, 90
Washington, 34, 88, 92, 93
waste, 76, 92
wastewater, 66
water, 8, 66
watershed, 52
welfare, 33, 81
wetlands, 66
White House, 93
WIA, 68
withdrawal, 79
witnesses, 24
workers, 68
Workforce Investment Act, 68, 91